T0183086

SpringerBriefs in Petroleum Geoscience & Engineering

Series Editors

Jebraeel Gholinezhad, School of Engineering, University of Portsmouth, Portsmouth, UK

Mark Bentley, AGR TRACS International Ltd, Aberdeen, UK

Lateef Akanji, Petroleum Engineering, University of Aberdeen, Aberdeen, UK

Khalik Mohamad Sabil, School of Energy, Geoscience, Infrastructure and Society, Heriot-Watt University, Edinburgh, UK

Susan Agar, Oil & Energy, Aramco Research Center, Houston, USA

Kenichi Soga, Department of Civil and Environmental Engineering, University of California, Berkeley, USA

A. A. Sulaimon, Department of Petroleum Engineering, Universiti Teknologi PETRONAS, Seri Iskandar, Malaysia

The SpringerBriefs series in Petroleum Geoscience & Engineering promotes and expedites the dissemination of substantive new research results, state-of-the-art subject reviews and tutorial overviews in the field of petroleum exploration, petroleum engineering and production technology. The subject focus is on upstream exploration and production, subsurface geoscience and engineering. These concise summaries (50-125 pages) will include cutting-edge research, analytical methods, advanced modelling techniques and practical applications. Coverage will extend to all theoretical and applied aspects of the field, including traditional drilling, shale-gas fracking, deepwater sedimentology, seismic exploration, pore-flow modelling and petroleum economics. Topics include but are not limited to:

- Petroleum Geology & Geophysics
- Exploration: Conventional and Unconventional
- Seismic Interpretation
- Formation Evaluation (well logging)
- Drilling and Completion
- Hydraulic Fracturing
- Geomechanics
- Reservoir Simulation and Modelling
- Flow in Porous Media: from nano- to field-scale
- Reservoir Engineering
- Production Engineering
- Well Engineering; Design, Decommissioning and Abandonment
- Petroleum Systems; Instrumentation and Control
- Flow Assurance, Mineral Scale & Hydrates
- Reservoir and Well Intervention
- Reservoir Stimulation
- Oilfield Chemistry
- Risk and Uncertainty
- Petroleum Economics and Energy Policy

Contributions to the series can be made by submitting a proposal to the responsible Springer contact, Anthony Doyle at anthony.doyle@springer.com.

More information about this series at http://www.springer.com/series/15391

Amin Taghavinejad · Mehdi Ostadhassan ·
Reza Daneshfar

Unconventional Reservoirs: Rate and Pressure Transient Analysis Techniques

A Reservoir Engineering Approach

 Springer

Amin Taghavinejad 🆔
Department of Petroleum Engineering
Amirkabir University of Technology
Tehran, Iran

Reza Daneshfar 🆔
Department of Petroleum Engineering
Petroleum University of Technology
Ahwaz, Iran

Mehdi Ostadhassan 🆔
Key Laboratory of Continental Shale
Hydrocarbon Accumulation and Efficient
Development, Ministry of Education
Northeast Petroleum University
Daqing, Heilongjiang, China

ISSN 2509-3126 ISSN 2509-3134 (electronic)
SpringerBriefs in Petroleum Geoscience & Engineering
ISBN 978-3-030-82836-3 ISBN 978-3-030-82837-0 (eBook)
https://doi.org/10.1007/978-3-030-82837-0

This Springer imprint is published by the registered company Springer Nature Switzerland AG
The registered company address is: Gewerbestrasse 11, 6330 Cham, Switzerland

This book is dedicated to:

Dr. Mohammad Sharifi,

professor of petroleum engineering at the Amirkabir University of Technology.

Acknowledgements

The authors would like to express their sincere appreciation to the Department of Petroleum Engineering at the Amirkabir University of Technology for their immense supports and feedback during the preparation of this book. Additionally, we'd like to thank the Key Laboratory of Continental Shale Accumulation and Efficient Development at the Northeast Petroleum University in Daqing China, for their continuous encouragement of our academic endeavors.

Last but not least Mehdi Ostadhassan would like to extend his sincere gratitude to his wife Bailey without whom and his full-time support, encouragement, and dedication his contribution to this book wouldn't have been possible. She was away when this book was being developed in a separate country to take care of our two beautiful sons, Elijah Ali and Jonas Reza.

Ultimately, we thank our friends, colleagues, and collaborators for their feedback, input, and scientific discussions which significantly improved the quality of this book.

Contents

Chapter 1
Unconventional Oil and Gas Reservoirs

1.1 Introduction to Unconventional Reservoirs

Unconventional resources or unconventional reservoirs (UR) refer to a category of underground hydrocarbon deposits which are different in operations and methods of recovery compared to conventional reservoirs. As a matter of fact, such discrepancies originate from their special geological, geochemical, and petrophysical characteristics. The most important feature of URs is their low-permeability nature. Also, some storage and flow mechanisms of these reservoirs are not common in conventional oil and gas deposits. Thus, it can be expected that URs require special studies in order to enhance their efficiency along with an economical production.

Considering the geology and structure of unconventional resources, it is significantly important that the majority of these reservoirs are seen as relatively deep, extensive, as well as greatly low-permeable porous rock layers. Hence most of these reservoirs are petroleum source rocks where hydrocarbon migration into upper permeable layers have not occurred, and subsequently conventional reservoir rocks cannot be identified distinctly. In fact, in this kind of reservoirs, the production operations take place directly from the source rocks. However, URs such as coal-bed methane and gas hydrates are exempted from this specific feature of unconventional resources; while, differences in physicochemical properties with conventional reservoirs, and the necessity of special production techniques will categorize them under URs. In this book our emphasis is mostly on shale and tight reservoirs all of which are often considered relatively deep source rocks with very low permeability and huge lateral extent. This being said, in the following section, important features of these reservoirs along with coal-bed methane—that in retrospect share similar characteristics with shale gas reservoirs—are discussed.

Schematic of a large-scale geological profile in Fig. 1.1, demonstrates structural geology, lateral extent, and geological sequence of these reservoirs as well as their differences from the conventional reservoirs in a glance. This figure indicates that the tight and shale oil/gas layers have an extensive distribution, accommodating for

A. Taghavinejad et al., *Unconventional Reservoirs: Rate and Pressure Transient Analysis Techniques*, SpringerBriefs in Petroleum Geoscience & Engineering, https://doi.org/10.1007/978-3-030-82837-0_1

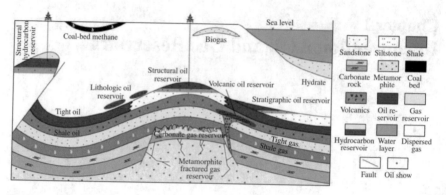

Fig. 1.1 Schematic geological profile of unconventional and conventional oil and gas accumulations [25]

large volumes of hydrocarbons compared to conventional reservoirs (accumulations stored in anticlines). The enormous amounts of hydrocarbons in such reservoirs reveals the importance of their exploitation if it is concluded that production would be economical.

1.2 Unconventional Gas Reservoirs

Unconventional gas reservoirs (UGRs) are mostly methane gas stored in tight to ultra-tight porous rock layers. However, other gas reservoirs such as shale gas condensate reservoirs has become the topic of recent studies by researchers and flow modelers in the past decade. Porosity and permeability in these reservoirs are considerably lower than conventional gas reservoirs. In other words, reservoir quality in such type in them is not suitable for natural production. Therefore, other techniques such as fracturing and horizontal well drilling would be necessary for their development [6]. In addition, an important feature of UGRs compared to conventional gas reservoirs is the difference in flow mechanisms than conventional gas reservoirs such as slip flow, Knudsen diffusion, surface diffusion along with other particular characteristics such as gas adsorption effect and stress dependence. It should be noted that most of these characteristics are observed in both shale gas reservoirs (SGRs) and coal bed methane (CBM).

Figure 1.2 is a conceptual triangular diagram of conventional and unconventional gas reservoirs illustrating an increase in reservoir volume, decrease in permeability and reservoir quality, and also advancement in technology that is required for production from unconventional gas reservoirs compared to the conventional ones.

In Fig. 1.3 a conceptual diagram is presented which depicts an increase in production from tight/shale gas in the United States which reflects the importance of their development and exploitation in the international energy market.

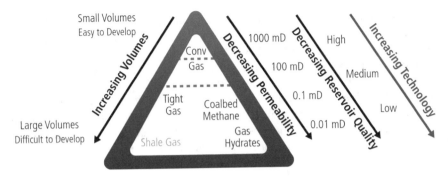

Fig. 1.2 Gas reservoirs triangle [1]

Fig. 1.3 US dry natural gas production by *sources* history and projection [7]

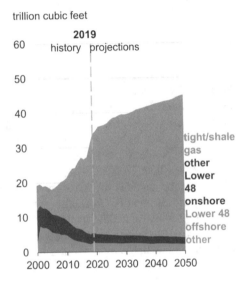

1.2.1 Tight Gas Reservoirs

Tight gas reservoirs (TGRs) generally have relatively low permeability (mostly less than 0.1 md) and porosity of less than 10%, which does not justify natural production from them via conventional single wells. TGRs are commonly found in the center and slopes of the sedimentary basins [25]. In these reservoirs, the distance of fluid migration from the source rock is short which is due to the proximity to the source rocks. There is no clear cap rock or barrier rocks, as well as sharp gas–water contact [25]. When the TGR is naturally fractured, majority of the flow happens under undamaged natural fractures crossing the wellbore [22].

The matrix consists of micropores, leading to high capillary pressure due to tiny pore openings. Moreover, because such tight reservoirs are usually hydrophilic in nature, capillary pressure leads to water absorption (imbibition) into the pores, and subsequently, high critical water saturation (S_{wc}), as well as irreducible water saturation (S_{wirr}) [3, 15]. Thus, the amount of gas saturation in these reservoirs might vary, although mostly reported less than 60% [25]. The initial water saturation (S_{wi} in tight reservoirs is innately equal to the critical water saturation, while occasionally it falls below critical saturation due to evaporation to the gas phase; when it surpasses the critical value of water saturation it would be because hydrocarbon trap has formed either during or after gas migration. It should be noted that the former scenario would increase the relative permeability of gas (something close to absolute permeability; and the later one will lead to a sharp diminish in relative permeability of the gas [4, 8].

1.2.2   Shale Gas Reservoirs

Shale gas reservoirs (SGRs) are porous shale rocks with nanometer-scale intergranular, intragranular, and organic pores storing natural gas. SGR rocks pore volume contains two types of gas—free and adsorbed gas. Unlike free gas which is a common feature of hydrocarbon storage between both conventional and unconventional gas reservoirs, adsorbed gas is a type of natural gas adsorbed on shale rocks in the form of a solid material. This form of natural gas could be produced from SGRs via desorption which takes place in low-pressure conditions [2]. Generally, permeability of these reservoirs is between 1 to 100 nano-Darcy, porosity less than 10%, and pore size is measured less than 2 nm to 2 μm [12]. Besides being ultra-tight and low-permeable, other common features of SGRs for economic development are high content of total organic carbon (TOC), moderate thermal maturity, high percentage of brittle minerals, presence of natural fractures, thickness, and relatively low burial depth [11, 25]. To make production from shale gas resources economically feasible, hydraulic fracturing is inevitable by making a highly-conductive environment comprised of transverse multistage fractures in a horizontal well near the wellbore area.

Shale gas is the product of high maturity kerogen generation of hydrocarbons under thermal pyrolysis as free and/or adsorbed gas. Shale gas is found in gentle slopes, areas under pressure, and margins of the basin. These gas layers are extremely thick, with low productions (about 10,000 m^3/d), while capable of 30–50 years of production without water injection [25]. Presence of nanopores in SGRs causes high capillary pressure while when the critical pressure and temperature declines, a distortion in the phase diagram happens resulting in capillary condensation and gas molecules slippage on the pores' walls (gas rarefaction). Moreover, because of intrinsic low permeability of the SGRs' matrix, Darcy flow will not be the sole flow mechanism governing in these reservoirs, and gas rarefaction as a result of fluid confinement in nanopores, plays an undeniable role in the transfer of fluid from the matrix to the micro and macro-fractures. Hydrocarbon-rich SGRs, unlike their counterparts, TGRs which are hydrophilic (water wet), are mostly oil-wet [12]. This means, in naturally fractured SGRs, free gas is stored in matrix pores and natural fractures, and the adsorbed gas is mostly found in the surface of the matrix grains, although the presence of adsorbed gas on natural fractures' surface as well, should be expected [18]. Numerous studies indicate that 5–30% of total gas production include desorbed gas, but the effect of gas desorption has been observed at the later stages of production from a well (low pressure conditions) [5, 16, 23]. Surface diffusion is another flow mechanism, that unlike conventional and TGRs, in SGRs is due to gas molecules diffusion on adsorption layer on the surface of organic/inorganic pores [21].

1.2.3 Coal-Bed Methane Reservoirs

Coalbed methane is another type of UGRs which is formed in coal layers, where the layer itself is both the source and a reservoir rock. In other words, there is no migration or short migration occurred in or from the source rock. As it is reported by [25], CBM reservoirs, contain more than 95% methane (CH_4), as well as heavy hydrocarbons such as ethane (C_2H_6) and propane (C_3H_8), and non-hydrocarbon including nitrogen (N_2) and carbon dioxide (CO_2). Also, porosity and permeability of CBM layers are usually less than 10% and 0.1 md, respectively.

Gas adsorption is a common phenomenon in CBM porous structure, alongside the free dry natural gas. The adsorbed gas quantity prevails the total gas content of the CBMs gas storage, however the dominant present form of gas in SGRs is the free gas [17]. Alike SGRs which are naturally fractured, CBM reservoir rocks are fractured as well, and the stored gas can be found in the matrix and cleats (fractures). It should be noted that although all dominant flow and storage mechanisms in CBM reservoirs are similar to SGRs but there is a discrepancy in how much each contributes to total gas flow. Therefore, in this book, SGRs flow modeling would be applicable to CBM reservoirs as well.

1.3 Unconventional Oil Reservoirs

Unconventional oil reservoirs (UORs), similar to UGRs, are low-permeable porous rocks but containing liquid petroleum fluid. Similarly, this kind of oil resources cannot produce without horizontal wells and hydraulic fracturing technologies in economic quantities. Therefore, recovery mechanisms in both UORs and UGRs would be the same, although flow mechanisms in such type of URs are not as complex as UGRs and pressure driven flow (Darcy's law) is the dominant driving force of fluid flow.

1.3.1 Tight Oil Reservoirs

The oil that co-exists in the low-permeable source rock, which has migrated a short distance, is known as tight oil which forms a tight oil reservoir (TOR) in porous media. These reservoirs are divided into tight sandstones and low permeable limestones based on the lithology of the formation. API degree of oil in this type of reservoirs reaches more than 40, and porosity and permeability of TORs would be generally less than 12% and 0.1 md, respectively [19, 25]. Production from TORs is accompanied by high pressure drawdowns which causes reservoir pressure to decline to lower than bubble-point and two-phase flow emerges [20]. Also, TORs mainly have very low oil recovery, considerable amount of oil remains in the pores, making enhanced oil recovery techniques vital for future development of these oil reservoirs [24].

Zou (2013) [25] has reported that tight sandstone oil reservoirs compared to conventional sandstone have relatively, higher water saturation, higher rock density, and higher capillary pressure. Furthermore, unlike typical sandstones that are not stress sensitive, they are strongly stress sensitive and pore sizes can vary with perturbation in the net stress. In addition, TORs with sandstone lithology often have abnormal and high formation pressure, unlike the conventional ones that are explored with normal or less than normal reservoir pressure. Figure 1.4 demonstrate the US crude oil production history and projection [7] which implies that tight oil reserves are and will be the main source of crude oil production in the US controlling international oil markets.

Fig. 1.4 US crude oil production by sources: history and projection [7]

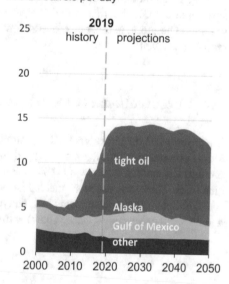

1.3.2 Shale Oil Reservoirs

Shale oil is referred to source rocks where migration of crude oil has not occurred. Shale oil reservoirs (SORs) are considered as oil deposits in a system of fractures and pores, where source rock and reservoir rock are the same. Based on the literature [10, 25], it is reported that in SORs TOC is more than 2%, porosity is less than 10%, and permeability is less than 0.1 md [13, 14]. In shale oils, hydrocarbon is stored in the nanoscale porous system of shale rocks in both free and adsorbed forms [10]. Similar to other unconventional petroleum resources, SORs should be drilled horizontally, and hydraulic fracturing should be implemented for economic production from them.

1.4 Unconventional Reservoirs Development

Following increase in tight and shale reservoirs development within two last decades specially in three leading countries (US, Canada, and China), technologies which made production from these low-quality reservoirs feasible becomes more important. Such technologies constitute horizontal drilling and multiple hydraulic fracturing, while their applicability and feasibility is proven in successful exploitation of US Barnett shale, or the Bakken [1]. This has led to emerging other advanced methods in petroleum industry for URs efficient production, such as micro-seismic monitoring, fracture modeling techniques, and fractured wells rejuvenation methods. A combination of these techniques is known as multi-stage fractured horizontal well (MFHW) planning which directly impacts the quality and quantity of production. Yet, it might not be adequate which makes implementation of other techniques necessary for optimum development. In this regard, one should not neglect, lithology determination, mineralogy analysis, total organic matter content (TOC), kerogen type, thermal maturity, prediction of the amount of adsorbed hydrocarbon, and quality of the rock, are important for better understanding of the URs and their hydraulic behavior. A schematic diagram of an MFHW is shown in Fig. 1.5 where M number of hydraulic fractures (HFs) with specific half-lengths on right and left wings (L_{FRi} and L_{FLi}, respectively) have crossed a horizontal well. Reservoir layer has a thickness h sandwiched between impermeable layers.

Based on what was said above, Ahmed and Meehan [1] argued that production from URs is strongly dependent on placement of horizontal wells and HF stages, mineralogy, and amount of the formation TOC. Also, they declared that development of the URs rely heavily on four main stages—exploration, appraisal, production, and rejuvenation. As it is shown in Fig. 1.6, 12 steps should be followed to guarantee success in development of URs.

This chart explains that building and modifying the reservoir model play the most important role in successful extraction of resources from URs. Ultimately, resource evaluation, lateral characterization, optimization of fracture design, and monitoring the production, pressure and rate transient analysis (PTA and RTA) techniques are

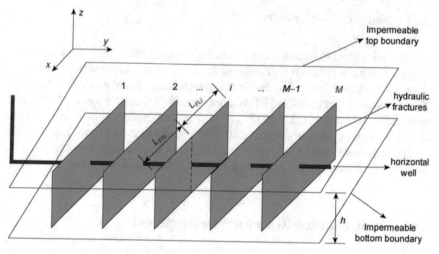

Fig. 1.5 Schematic view of an MFHW [9]

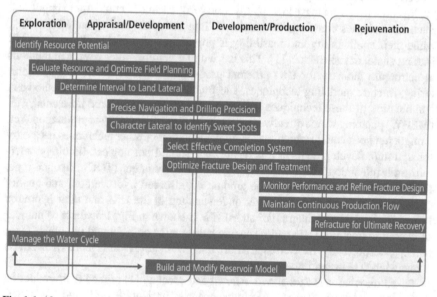

Fig. 1.6 12 steps to success in URs development—proposed based on Baker Hughes data [1]

all useful methods in URs development that should benefit from model building and
iterative modifications.

References

1. Ahmed U, Meehand D (2016) Unconventional oil and gas resources. CRC Press, Boca Raton, Florida, Explotation and Development
2. Bello RO, Wattenbarger RA (2009) Rate transient analysis in shale gas reservoirs with transient linear behavior. Texas A&M University
3. Bennion DB, Thomas FB (2005) Formation damage issues impacting the productivity of low permeability, low initial water saturation gas producing formations
4. Bennion DB, Thomas FB, Bietz RF (1996) Low permeability gas reservoirs: problems, opportunities and solutions for drilling, completion, stimulation and production. In: SPE gas technology symposium. Society of Petroleum Engineers
5. Cipolla CL, Lolon EP, Erdle JC, Rubin B (2010) Reservoir modeling in shale-gas reservoirs. SPE Reserv Eval Eng 13:638–653
6. Dahim S, Taghavinejad A, Razghandi M et al (2020) Pressure and rate transient modeling of multi fractured horizontal wells in shale gas condensate reservoirs. J Pet Sci Eng 185:106566. https://doi.org/10.1016/j.petrol.2019.106566
7. EIA (2020) Annual Energy Outlook 2020. United States
8. Gonfalini M (2005) Formation evaluation challenges in unconventional tight hydrocarbon reservoirs. SPE Ital Sect slide# 15 25
9. Gu D, Ding D, Gao Z et al (2017) Pressure transient analysis of multiple fractured horizontal wells in naturally fractured unconventional reservoirs based on fractal theory and fractional calculus. Petroleum 3:326–339
10. Jiang Z, Zhang W, Liang C et al (2016) Basic characteristics and evaluation of shale oil reservoirs. Pet Res. https://doi.org/10.1016/S2096-2495(17)30039-X
11. Kuuskraa VA (2004) Natural Gas Resources, Unconventional. In: Cleveland CJ (ed) Encyclopedia of energy. Elsevier Science, pp 257–272
12. Lee KS, Kim TH (2016) Integrative Understanding of Shale Gas Reservoirs. Springer
13. Liu K, Ostadhassan M (2017) Quantification of the microstructures of Bakken shale reservoirs using multi-fractal and lacunarity analysis. J Nat Gas Sci Eng 39:62–71. https://doi.org/10.1016/j.jngse.2017.01.035
14. Liu K, Ostadhassan M (2017) Microstructural and geomechanical analysis of Bakken shale at nanoscale. J Pet Sci Eng 153:133–144. https://doi.org/10.1016/j.petrol.2017.03.039
15. Mahadevan J, Sharma MM, Yortsos YC (2007) Capillary wicking in gas wells. SPE J 12:429–437
16. Mengal SA, Wattenbarger RA (2011) Accounting for adsorbed gas in shale gas reservoirs. In: SPE middle east oil and gas show and conference. Society of Petroleum Engineers
17. Sharma S, Saxena A, Saxena N (2019) Unconventional Resources in India: The Way Ahead. Springer Nature, Switzerland
18. Song B, Economides MJ, Ehlig-Economides CA (2011) Design of multiple transverse fracture horizontal wells in shale gas reservoirs. In: SPE Hydraulic fracturing technology conference.
19. Su Y, Zha M, Liu K et al (2021) Characterization of pore structures and implications for flow transport property of tight reservoirs: a case study of the Lucaogou Formation, Jimsar Sag, Junggar Basin. Northwestern China. Energies 14:1251. https://doi.org/10.3390/en14051251
20. Tabatabaie SH, Pooladi-Darvish M (2017) Multiphase linear flow in tight oil reservoirs. SPE Reserv Eval Eng
21. Taghavinejad A, Sharifi M, Heidaryan E et al (2020) Flow modeling in shale gas reservoirs: a comprehensive review. J Nat Gas Sci Eng 83:103535. https://doi.org/10.1016/j.jngse.2020.103535
22. Teufel LW, Chen H-Y, Engler TW, Hart B (2004) Optimization of infill drilling in naturally-fractured tight-gas Reservoirs. New Mexico Institute of Mining and Technology (US)

23. Thompson JM, Okouma Mangha V, Anderson DM (2011) Improved shale gas production forecasting using a simplified analytical method-a marcellus case study. In: North American Unconventional Gas Conference and Exhibition. Society of Petroleum Engineers
24. Yu W, Sepehrnoori K (2019) Introduction of Shale Gas and Tight Oil Reservoirs. In: Shale gas and tight oil reservoir simulation. Gulf Professional Publishing
25. Zou C (2013) Unconventional Petroleum Geology. In: Elsevier

Chapter 2
Unconventional Reservoir Engineering

2.1 Fluid Flow in Reservoirs

The most fundamental concept in reservoir engineering is the fluid flow. Considering reservoirs as porous materials (rock) that are saturated with hydrocarbons (fluid), flow in the reservoir can be described through general understanding of dynamics of hydrocarbons flow in porous media.

Describing how a fluid can move through porous structure in the reservoir rock and reaches the wellbore would be necessary for a reservoir engineer to evaluate the recovery and also productivity/injectivity of the production/injection wells. In this regard, geological heterogeneities and other complexities such as existence of natural fractures (NFs) may complicate flow modeling in porous media in the reservoir. In addition, in unconventional reservoirs (URs) that are mainly low-permeability rock layers, stimulation by hydraulic fracturing enhances the reservoir rock hydraulic conductivity of the producing/injecting fluid. This will add another layer of complexity to flow modeling efforts which will be discussed later when fundamentals of flow modeling are first discussed.

2.1.1 Hydraulically Fractured Wells

Since URs need more contact area of the well and the reservoir, horizontal well drilling is a necessity in development of these kind of petroleum plays. Also, to create more permeable pathways in the reservoir, taking advantage of multi-stage fractured horizontal wells (MFHWs) has been confirmed to be an effective method in exploitation of these low-permeability reservoirs.

In Fig. 2.1, schematic view of an MFHW with 4 frac stages are shown where the reservoir volume is divided into two separate volumes after hydraulic fracturing: stimulated reservoir volume (SRV) and unstimulated reservoir volume (USRV).

© The Author(s), under exclusive license to Springer Nature Switzerland AG 2022
A. Taghavinejad et al., *Unconventional Reservoirs: Rate and Pressure Transient Analysis Techniques*, SpringerBriefs in Petroleum Geoscience & Engineering,
https://doi.org/10.1007/978-3-030-82837-0_2

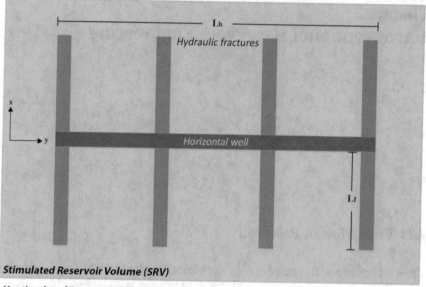

Fig. 2.1 Schematic of a 4-stage MFHW dividing reservoir volume to SRV and USRV

UR properties in the USRV remains unchanged while in SRV petrophysical properties are stimulated, and are expected to be improved to facilitate economic production. However, considering significant heterogeneity in such systems, reservoir properties between individual HFs in the SRV won't be similar or even equally affected by stimulation. In addition, in spite of desired outcomes for horizontal well drilling and hydraulic fracturing that is expected following modeling efforts, some properties such as frac half-length (L_f), frac height, frac width, as well as horizontal well length (L_h), won't be the only controlling factors with proven impacts. Therefore, to obtain better results in the field, further studies beyond these properties would be required. plays which plays an important role in unconventional reservoir engineering (URE) that would guarantee well performance. This can be done via certain methods like well test interpretation, the analysis of the production decline rate, or history matching of the production data.

2.1.2 Naturally Fractured Reservoirs

Naturally fractured reservoirs (NFRs) contain fissures dividing porous rocks into two distinct portions: matrix and fractures. Considering that fact that unconventional reservoirs are generally naturally fractured, and can open by stimulation operations.

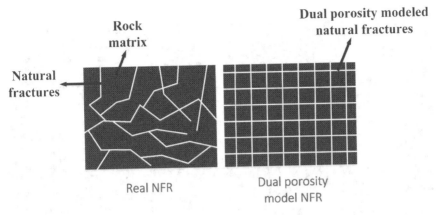

Fig. 2.2 Schematic diagram of real and dual-porosity model of a naturally fractured reservoir [10, 31]

Furthermore, artificially induced fractures through stimulation will play a similar role to NFs since they would equally contribute to the fluid flow.

A simplified method for taking NFs into account in flow modeling in the reservoir is proposed by Warren-Root known as the dual-porosity model [38]. This model is based on the pseudo-steady state flow regime of fluids from the matrix to the fracture where, the matrix is considered as cubes separated by fractures. Schematic diagram of a dual-porosity NFR model compared with a real NFR is shown in Fig. 2.2.

In addition to the dual-porosity model, triple-porosity model is another alternative one that can be applicable to shale and tight plays. This model considers two kinds of matrix with distinct petrophysical properties. Where both types of matrix media own inter-porosity flows between the matrix and the fractured media. Hence, the effects of dual-porosity model in NFRs under triple-porosity model can be observed in two distinct ways: one due to inter-porosity between matrix No.1 and the fracture and the other one due to matrix No.2 and the fracture inter-porosity flow. Figure 2.3 illustrates a comparative schematic of a dual-porosity (Fig. 2.3a) and a triple-porosity (Fig. 2.3b) model.

Besides the complexities that is raised in fluid flow modeling due to the existence of HFs and NFs in a reservoir, gas adsorption, gas rarefaction, and sensitivity of rock matrix and NFs to overburden pressure during production will add sophistication in modeling of tight and shale reservoirs' fluid flow which will be briefly outlined.

2.1.3   Fluid Flow in Porous Media of URs

Fluid flow in URs' porous media differs from conventional reservoirs to some extent in areas such as: gas adsorption and rarefaction in submicron pores and their stress dependency which are elaborated prior to explaining flow modeling in URs.

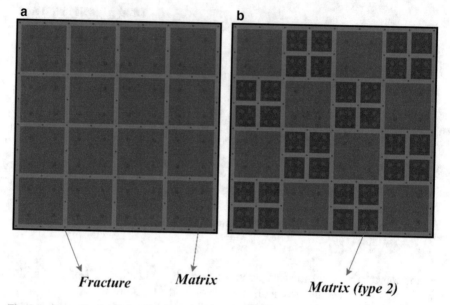

Fig. 2.3 Comparison of 2-D schematics of dual-porosity (**a**) and triple-porosity (**b**) NFR models [32]

2.1.3.1 Gas and Oil Sorption

Gas Sorption

In shale gas reservoirs (SGRs), in addition to free gas in the pores, another form of stored hydrocarbon known as adsorbed gas exists which can be found adsorbed on organic/inorganic matrix parts [9], or rarely on the surface of the fracture [29]. For the adsorbed fluid to flow, it has to be desorbed from grains to flow along the free gas. Figure 2.4 demonstrate the schematics of gas shale matrix pores and fractures containing fluids where free gas can be stored and flow in both matrix and fractured media, whereas the adsorbed/desorbed gas is only found in the matrix.

Based on what was said above, gas adsorption/desorption isotherms enable us to model quantities of the gas in the pores. Taghavinejad et al. [33] reviewed all applicable and proposed isotherms, and argued that some are more prominent than others to make them more popular as summarized in Table 2.1. Langmuir model is the one that is most commonly used which accounts for monolayer gas sorption while BET model takes into account multilayer sorption phenomena. D-A method is known as pore-filling model, and considers van Der Waals forces as the driving force for gas-surface attraction. Also, Toth model is an extended form of Langmuir which is more capable of handling experimental data. Based on the study by Zou et al. (2018), pore-filling and monolayer adsorption phenomena are dominant in

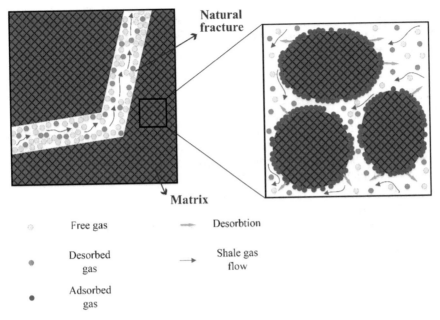

Fig. 2.4 Schematic view of stored and the flowing gas in SGRs [10]

Table 2.1 Common isotherm models for shale gas sorption

Isotherm model	Formulation
Langmuir [16]	$V = \frac{V_L P}{P + P_L}$
BET [5]	$V = \frac{V_m C P}{(P_0 - P)\left[1 + \frac{(C-1)P}{P_0}\right]}$
D-A [12]	$V = V_m \exp\left\{-\left[D\ln\left(\frac{P_0}{P}\right)\right]^n\right\}$
Toth [3]	$V = \frac{V_L b P}{\left[1 + (bP)^k\right]^{(1/k)}}$

micropores and mesopores of shales, respectively. Hence, they pointed out that if D-A is coupled with Langmuir, sorption phenomena in shale reservoirs with complex pore size distribution (PSD) can better be represented.

In Table 2.1, V is the adsorption capacity or volume (m³/kg), and P is pressure (Pa). In Langmuir and Toth models, V_L denotes the Langmuir volume (m³/kg), P_L denotes the Langmuir pressure (Pa), k is the dimensionless Henry's constant, and b is the adsorption affinity ($b = 1/P_L$). Considering the BET model, V_m is the maximum adsorption capacity when surface is covered by monolayer molecules of gas, C denotes the net heat adsorption constant, and P_0 is the saturation pressure of the gas. Constant C can be written as $C = \exp\left(\frac{E_1 - E_L}{RT}\right)$ where E_1 is the adsorption heat of the first layer, E_L is the adsorption heat of the second and next layers, R is the real gas constant, and T is the temperature. Finally, in D-A isotherm, n is a tuning exponent and D is the empirical binary interaction coefficient. Coefficient D is defined as $D = \frac{RT}{\beta E}$ where β is the affinity coefficient of the adsorbed gas and E is the characteristic energy.

Oil Sorption

In addition to gas sorption in shale reservoirs, oil shales can adsorb on or desorb from the grains in shale plays as well. Research in this area is limited but rapidly expanding. Some scholars suggest molecular dynamics (MD) simulation would be a reliable tool compared to mathematical modeling to understand oil sorption phenomena [19, 37, 39] while computation costs cannot be neglected.

As it is proposed by Li et al. (2018) [18], adsorbed and free amounts of shale oils per rock mass unit can be expressed by the following expressions, respectively:

$$V_a = k d_m S_t h n \tag{2.1}$$

$$V_f = V_o - k d_m S_t h n \tag{2.2}$$

where V_a and V_f are adsorbed and free amounts of shale oil (m³/kg), V_o is the volume of open rock pores containing oil (m³/kg), d_m is the mean pore diameter (m), S_t is specific surface area open pores containing oil (m²/kg), h is the mean monolayer adsorption thickness (m), and n is the mean adsorption layer number. Also, k is a coefficient (m^{-1}) which is defined as:

$$k = \frac{\lambda}{F h n} \tag{2.3}$$

where λ is the ratio of adsorbed phase to the total pore volume (dimensionless), and F denotes the morphological factor (dimensionless), equal to 6, 4, and 2 for spherical, columnar, and slit-shaped pores, respectively.

Coefficient k, is theoretically between zero and $1/d_m$. When k is equal to zero, it corresponds to $P/P_0 = 0$, meaning adsorption has not occurred between oil and the rock pore surface. When $0 < k < 1/d_m$, it defines $0 < P/P_0 < 1$, and oil adsorbs on the pore surfaces (unsaturated adsorption). Ultimately, k equal to $1/d_m$ denotes $P/P_0 = 1$, the maximum adsorption on the pore surfaces (saturated adsorption).

2.1.3.2 Gas Rarefaction

The flow of gas in micro-/nano-size pores in a media causes the gas molecules to slip on the surface, thus flow will not be solely due to pressure difference. This means continuity assumption will not be valid anymore. It is known that Darcy's law represents pressure-driven flow while flow due to gas slippage on the pore surface should be added to the classical Darcy's law formulation. This is common in tight/shale gas reservoirs when average pore size is in the scale of micro-/nano-meter.

The effect of gas slippage was first introduced in petroleum engineering by Klinkenberg [14] as below:

$$k_g = k_0\left(1 + \frac{b}{P_m}\right) \tag{2.4}$$

where k_g is gas permeability, k_0 is intrinsic permeability of the rock, b is slippage factor, and P_m is the average pressure. It is inferred that gas permeability increases as the average pressure decreases, thus when pressure is increased, the term b/P_m will become negligible and $k_g \approx k_0$.

Afterward, gas rarefaction was introduced as a flow mechanism which differentiates gas flow in conventional reservoirs. Gas rarefaction is the general state of gas slippage—depending on the pressure—when flow mechanism of the gas can be either slip, transitional, or Knudsen flows. Here, Knudsen number (Kn) is defined as [22]:

$$Kn = \frac{\lambda}{L_c} \tag{2.5}$$

where λ denotes the average mean free path (m), and L_c denotes the characteristic length of the porous medium (m) and average mean free path is defined as:

$$\lambda = \frac{\mu}{P}\sqrt{\frac{\pi RT}{2M}} \tag{2.6}$$

where μ is gas viscosity (Pa s), and M is the molecular weight of the gas (kg/mol).

Different ranges of Kn would determine the flow mechanisms of the gas in micro-/nano-channels, regardless if the medium is porous or not [30]. Sunden and Fu [30] proposed when Kn < 0.001 rarefaction isn't expected and continuum flow is dominant; when 0.001 < Kn < 0.1 gas slippage occurs, continuum flow is no longer dominant, and slip flow would be the main driving force; when 0.1 < Kn < 10 a transitional flow is the main flow mechanism of gas molecules; and when Kn > 10 free molecular flow (Knudsen diffusion) happens. In lower Kn values (continuum and slip flows), molecule–molecule interactions are dominant, while in higher Kn values (free molecular flow) molecule-wall interactions are the main molecular collisions in the micro-/nano-size pores.

One of the most applicable models for permeability evaluation under gas rarefaction effect is the Javadpour (2009) [13] model defined by Eq. 2.7 where both slip flow and Knudsen diffusion are considered in the apparent permeability estimation.

$$k_a = \left(F_{slip} + \frac{b_a}{P} \right) k_0 \tag{2.7}$$

where k_a is apparent permeability (m^2), F_{slip} is slip flow dimensionless correction factor, b_a is the Knudsen diffusion factor (Pa), and P is pressure (Pa). It is clear that in the case of continuum flow, F_{slip} is equal to unity, b_a is zero, and apparent permeability will be equal to intrinsic permeability of the rock. F_{slip} and b_a are represented by Eqs. 2.8 and 2.9, respectively:

$$F_{slip} = \left(\frac{8\pi RT}{M} \right)^{0.5} \frac{\mu}{r} \left(\frac{2}{\sigma} - 1 \right) \tag{2.8}$$

$$b_a = \frac{P c_g \mu D_K}{k_0} \tag{2.9}$$

where σ is tangential momentum accommodation coefficient (TMAC), c_g is isothermal compressibility coefficient of gas (Pa^{-1}), and D_K is Knudsen diffusivity (m^2/s) and is written as:

$$D_K = \frac{d_p}{3} \sqrt{\frac{8RT}{\pi M}} \tag{2.10}$$

where d_p is pore diameter.

2.1.3.3 Stress Dependence

The compaction of rocks due to the change in net effective stress on bed-rocks alters the mean pore sizes. In this regard, as net effective stress increases, the porosity and permeability of the rock will be decreased [17]. It is reported by Ahmed and Meehan [1], that stress sensitivity of the rocks and geomechanical attributes in development of URs are more complicated than conventional resources due to transverse anisotropy, existence of NFs, and difficulty for measurement of pore and fracture closure pressures. The anisotropy is usually because of their geological characteristic (i.e., lamination for shales), brittleness of minerals, and distribution of organic matter within the rock structure.

Various models are proposed to define how permeability and porosity would change versus net confining pressure. Nonetheless, the most prominent ones are introduced in Table 2.2 and addressed in details by Taghavinejad et al. [33]. In this table, φ_σ and φ_0 are the stress dependent porosity and intrinsic porosity under initial conditions of the rock, respectively. Also, k_σ and k_0 are stress dependent permeability and intrinsic permeability under initial conditions of the rock, respectively.

Table 2.2 Common isotherm models for shale gas sorption

Stress sensitivity model	Formulation	Explanations
[24]	$\dfrac{\varphi_\sigma}{\varphi_0} = 1 - \left(1 + \dfrac{2}{\varphi_0}\right)\Delta\varepsilon_s$ $\dfrac{k_\sigma}{k_0} = \left(\dfrac{\varphi_\sigma}{\varphi_0}\right)^3$	– Developed for cleats/fractures – $\Delta\varepsilon_s$: sorption-induced volumetric strain; $\Delta\varepsilon_s = \varepsilon_1\left(\dfrac{P}{P+P_L} - \dfrac{P_0}{P_0+P_L}\right)$ – ε_1: matrix shrinkage; $\varepsilon_1 = s_m V_L$
[20]	$\dfrac{\varphi_\sigma}{\varphi_0} = 1 - \dfrac{c_m}{\varphi_0}\Delta P + \dfrac{2}{3\varphi_0}\dfrac{2v-1}{1-v}\Delta\varepsilon_s$ $\dfrac{k_\sigma}{k_0} = \left(\dfrac{\varphi_\sigma}{\varphi_0}\right)^3$	– Developed for naturally fractured rock matrices – c_m: matrix compressibility coefficient (Pa^{-1}); $c_m = \dfrac{(1+v)(1-2v)}{E(1-v)} -$ $\left[\dfrac{2}{3}\left(\dfrac{2v-1}{1-v}\right) + f_p\right]c_{grain}$ – c_{grain}: grain compressibility coefficient (Pa^{-1}); $c_{grain} = \dfrac{3(1-2v)}{E}$ – v: Poisson's ratio – E: denotes Young's modulus (Pa) – f_p: a constant between 0 and 1
[25]	$\sigma - \sigma_0 = -\dfrac{v}{1-v}(\Delta P) + \dfrac{E}{1-v}\Delta\varepsilon_s$ $\dfrac{k_\sigma}{k_0} = e^{-3c_f(\sigma-\sigma_0)}$	– Developed for cleats/fractures c_f: fracture/cleat compressibility (Pa^{-1})

In addition to the sorption-induced models summarized in Table 2.2, several mathematical models have been introduced as well. These models can be applied to both stress sensitive permeability and/or porosity. Based on the study by Raghavan and Chin [21], exponential and linear stress sensitivity models are defined by Eqs. 2.11 and 2.12 respectively.

$$\frac{\Phi_\sigma}{\Phi_0} = e^{-a\Delta P} \tag{2.11}$$

$$\frac{\Phi_\sigma}{\Phi_0} = 1 - b\Delta P \tag{2.12}$$

where Φ can be either permeability or porosity, and ΔP denotes pressure difference between effective and initial pressure values ($\Delta P = P-P_0$). Also, a and b are tuning coefficients are determined using experimental data.

Furthermore, based on Dong et al. (2010) [11], power-law stress sensitivity model is introduced as:

$$\frac{\Phi_\sigma}{\Phi_0} = \left(\frac{P}{P_0}\right)^{-c} \tag{2.13}$$

where c is the tuning coefficient.

2.1.3.4 Flow Modeling in URs

Fundamental flow equations in porous media are represented by partial differential equations (PDEs) that include the physiques of flow containing pressure variations extended over time. Thus, these flow equations are PDEs including derivatives for both reservoir dimension (r) and time (t). Alike flow modeling formulae for NFRs, governing flow equations of SGRs have to be expressed for both fracture and matrix media.

Continuity equation—regarding the conservation of momentum—for SGR fracture medium in spherical coordinates can be expressed as:

$$\frac{1}{r^2}\frac{\partial}{\partial r}\left(r^2\rho_f u\right) = \frac{\partial}{\partial t}(\rho_f\varphi_f) - q_m \tag{2.14}$$

where ρ_f is gas density in fractures, u is gas velocity, φ_f is porosity of the fracture, and q_m is the matrix-fracture transfer rate (kg/m^3/s).

According to the Warren-Root dual-porosity model, matrix-fracture transfer rate is:

$$q_m = \alpha\frac{k_m}{\mu}(\rho_m P_m - \rho_f P_f) \tag{2.15}$$

where α denotes the shape factor (m^{-2}), k_m is matrix permeability, μ is gas viscosity, ρ_f is gas density in matrix, P_m is gas pressure in matrix, and P_f is fracture pressure in matrix.

Using Eqs. 2.14 and 2.15 and definition of dimensionless and inter-porosity parameters (Table 2.3) plus following detailed derivation by Zhao et al. [40], flow equations in fracture and matrix in terms of pseudo-pressure will have the following forms, respectively:

$$\frac{1}{r_D^2}\frac{\partial}{\partial r_D}\left(r_D^2\frac{\partial m(P_f)}{\partial r_D}\right) = \omega_f\frac{\partial m(P_f)}{\partial t_D} - \lambda(m(P_m) - m(P_f)) \tag{2.16}$$

$$-\lambda(m(P_m) - m(P_f)) = (1 - \omega_f)\frac{\partial m(P_m)}{\partial t_D} + \omega_d\frac{\partial m(P_m)}{\partial t_D} \tag{2.17}$$

In Table 2.3, h is formation thickness, k_f is permeability of natural fracture, L is horizontal well half length, h_m is length of matrix block, and n is the number

Table 2.3 Definition of dimensionless and inter-porosity parameters

$r_D = \frac{r}{L}$	Dimensionless radius
$t_D = \frac{k_f}{(\varphi_f c_{tf} + \varphi_m c_{tm})\mu L^2} t$	Dimensionless time
$m_D(P) = 86400 \frac{\pi k_f h T_{sc}}{P_{sc} q_{sc} T} \Delta m(P)$	Dimensionless pseudo-pressure
$c_d = \frac{P_{sc} T (1 - \varphi_m - \varphi_f)\bar{Z}_m}{\varphi_m T_{sc}} \frac{V_L}{(P_L + \bar{P}_m)^2}$	Desorption compressibility
$\lambda = \alpha \frac{k_m}{k_f} L^2, \ \alpha = \frac{4n(n+2)}{h_m^2}$	Inter-porosity flow coefficient
$\omega_f = \frac{c_{tf}\varphi_f}{c_{tf}\varphi_f + c_{tm}\varphi_m}$	Fracture storativity ratio
$\omega_m = \frac{c_{tm}\varphi_m}{c_{tf}\varphi_f + c_{tm}\varphi_m} = 1 - \omega_f$	Matrix storativity ratio
$\omega_d = \frac{c_d\varphi_m}{c_{tf}\varphi_f + c_{tm}\varphi_m}$	Desorption storativity ratio

of natural fractures sets perpendicular to each other. Also, $\Delta m(P)$ is the difference between the initial and current pseudo-pressure values $(m(P_i) - m(P))$.

Since analytical solutions for the governing flow equations have to be applied in other realms such as pressure transient analysis (PTA) and rate transient analysis (RTA), Laplace transform would be the main approach to solve flow equations analytically. Equations 2.18 and 2.19 indicate the Laplace space flow equations in fracture and matrix, respectively, in terms of pseudo-pressure difference.

$$\frac{1}{r_D^2} \frac{\partial}{\partial r_D} \left(r_D^2 \frac{\partial \Delta \bar{m}(P_f)}{\partial r_D} \right) = \omega_f s \Delta \bar{m}(P_f) - \lambda(\Delta \bar{m}(P_m) - \Delta \bar{m}(P_f)) \quad (2.18)$$

$$-\lambda(\Delta \bar{m}(P_m) - \Delta \bar{m}(P_f)) = (1 - \omega_f)s \Delta \bar{m}(P_m) + \omega_d s \Delta \bar{m}(P_m) \quad (2.19)$$

where s is the Laplace transform parameter.

By rearranging Eq. 2.19 and inputting it to Eq. 2.18, final form of flow diffusivity equation in fracture media of SGRs will result as Eq. 2.20 and 2.21.

$$\frac{1}{r_D^2} \frac{\partial}{\partial r_D} \left(r_D^2 \frac{\partial \Delta \bar{m}(P_f)}{\partial r_D} \right) = s f(s) \Delta \bar{m}(P_f) \quad (2.20)$$

$$f(s) = \frac{\lambda(1 + \omega_d) + \omega_f(1 - \omega_f + \omega_d)s}{\lambda + (1 - \omega_f + \omega_d)s} \quad (2.21)$$

Abovementioned dimensionless equation of gas flow in SGRs should be solved using analytical, semi-analytical, or numerical solutions for further applications in RTA or PTA. Semi-analytical solution (analytical solution coupled with numerical Laplace inversion) of this equation and its applications will be discussed in Chap. 3 for PTA and Chap. 4 for RTA analyses.

It is noteworthy that the difference between $f(s)$ term in shale reservoir and conventional or tight gas reservoirs is in desorption storativity ratio ω_d and consequently presence of isothermal compressibility of desorption c_d in formulations. If these parameters are set to zero, flow equations of conventional/tight gas reservoirs will be yielded. In addition for oil reservoirs, a similar formulation can be derived following a similar procedure [32]. Equations 2.22 and 2.23 explain fracture medium flow equations in naturally fractured oil reservoirs which is applicable to both tight and shale oil reservoirs.

$$\frac{1}{r_D} \frac{\partial}{\partial r_D} \left(r_D \frac{\partial \overline{P}_{fD}}{\partial r_D} \right) = sf(s)\overline{P}_{fD} \tag{2.22}$$

$$f(s) = \omega_f + \frac{\lambda(1 - \omega_f)}{\lambda + (1 - \omega_f)s} \tag{2.23}$$

In these equations, $\overline{P}$ denotes the Laplace space oil pressure. Except dimensionless pressure, all dimensionless and inter-porosity flow parameters are similar to those defined in Table 2.3. Dimensionless pressure for oil reservoirs is formatted as Eq. 2.24. Detailed derivation can be found in [32].

$$P_D = \frac{k_f h}{q \mu B_o} \Delta P \tag{2.24}$$

where B_o is the formation volume factor (reservoir volume to standard condition volume ratio for oil) and q is oil volume flow rate in standard conditions.

2.2 Unconventional Reservoir Engineering Workflow

Reservoir engineering is referred to: (1) characterization of reservoir rock and fluid properties, (2) estimating the amount of the in-situ hydrocarbons, (3) controlling the reservoir recovery during the life of the reservoir, (4) production forecasting, and (5) economic evaluation and field development planning. Reservoir engineers would need a variety of tools; field data, experimental analyses, reservoir simulation, RTA/PTA techniques, etc. to fulfill their goals. In this regard, RTA and PTA methods are vital for reservoir engineers—that are also known as production decline curve analysis and well test analysis in classical terminology respectively. These methods directly help reservoir engineers to achieve items 1, 2, and 4 of reservoir engineering, and also indirectly support their efforts in items 3 and 5.

In general, a set of activities in URE can be considered through a workflow. This workflow, which is based on characterization of the reservoir and hydraulic fractures (HFs), reservoir and HFs modeling, economic analysis, and development planning, clarify the tasks that a reservoir engineer should undertake to achieve objectives

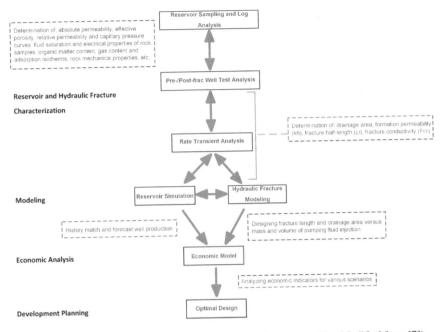

Fig. 2.5 Workflow for unconventional reservoir engineering proposed by (Modified from [7])

defined in URE. This workflow, which has been introduced initially by Clarkson et al. [7] for unconventional gas reservoirs (UGRs), describes the steps that can be taken to develop unconventional oil and gas fields. The flowchart of this workflow is shown in Fig. 2.5

Considering this workflow, it is understood that knowing reservoir and HF properties would play a key role in URs engineering and development. Table 2.4 summarizes the most important properties of the reservoir and HFs, required for reservoir performance evaluation, and the source of acquiring this information [1, 7, 28].

Various methods for reservoir/HF properties characterization are briefly reviewed in the following sections while their relevance to PTA and RTA will be explained in detail in Chaps. 3 and 4, respectively.

2.2.1   Reservoir and Hydraulic Fractures Characterization

At this stage of URE, a collection of properties and characteristics of the reservoir and well system have to be acquired. As it is summarized in Table 2.4, a variety of data should be obtained from different sources such as core, well logging, well testing, and production analyses. In the following section, we will introduce and explain each of these sources and how they can provide reservoir engineers, the data they might need.

Table 2.4 Definition of dimensionless and inter-porosity parameters

Reservoir properties

Inert gas expansion, mercury injection capillary pressure (MICP), nuclear magnetic resonance (NMR), core analysis calibrated log-analysis, Gas Research Institute (GRI) crushed core analysis, and Focused Ion Beam—Scanning Electron Microscopy (FIB-SEM) tomography	Porosity
Pressure/pulse decay core analysis, MICP, pre-/post-frac injection/falloff test (IFOT), diagnostic fracture injection test (DFIT), post-frac build-up test (BUT), RTA, simulation history matching (HM), and GRI crushed core analysis	Permeability
Mud logging, and Laboratory PVT analysis of oil, water, and gas phases	Fluid properties
Open-hole logging, production logging, frac job, and IFOT	Temperature
IFOT, perforation inflow diagnostic (PID), and perforation inflow test analysis (PITA)	Pore pressure
Core extraction analysis (Dean Stark or Retort methods), capillary pressure method, log-analysis, and GRI crushed core analysis	Water saturation
Leco-TOC and Roc-Eval	Total organic carbon (TOC)
IFOT, frac job, log-analysis (dipole sonic imager, DSI)	Fracture and closure stress
Canister desorption and adsorption isotherms (i.e., Langmuir isotherm)	Adsorbed gas capacity
Hydraulic fracture properties	
Post-frac net-pressure analysis, post-fracture BUT, and RTA	Propped HF length and conductivity
Microseismic, tiltmeter survey, and 4D seismic	Created HF length, height, and complexity

2.2.1.1 Core Analysis

Core analysis of unconventional reservoirs encompasses laboratory study of petro-physical and geochemical properties of URs' samples, which is performed on whole-cores, core-plugs, cuttings, or outcrops as various scales of investigation. Evidently, dimensional scale reduces in size from the whole-core and core-plugs to cuttings. As it is reported by Ahmed and Meehan [1], length of a whole-core is approximately 30 to 60 feet with a diameter of 2 to 6 inches while 4 to 5.3 inches is reported as the most common diameters. Also, for the sidewall core-plugs, 5.1 inches in diameter and 5.3 inches in length are reported as the approximate dimension. In general, whole-cores are better representative of the reservoir particularly in less heterogeneous ones, compared to other tangible specimens. However, such they are more sensitive to mud permeation, and their washing, cleaning, and further employment is more challenging compared to the others. In addition, core-plugs, since they are less sensitive to being contaminated by drilling mud compared to the whole-cores, have their particular benefits and accurate determination of the properties of lithofacies can be done using this sample size more effectively [7].

Another issue that must be addressed in the concerning core analysis is whether to grind the samples for measuring their properties or not while both approaches have their own advantages and disadvantages [28]. Since reservoir coring is accompanied by creating artificially induced fissures due alteration of stresses, some suggests the samples should be ground in order to mitigate the effects caused by microfractures. This could have two main problems; (1) grinding rock samples is an exhaustive task; and (2) permeability of the samples, which is a pore-size dependent parameter, would change as a result of grinding [8]. On the contrary since using core-plugs would increase the probability of having induced fissures in the rock which can impact certain properties, it is recommended that such experiments performed under confining pressures to keep microfractures closed under pressures similar to the reservoir conditions [7].

2.2.1.2 Well Log Analysis

Well logging refers to running tools in the well, and measuring petrophysical properties simultaneously. For example, neutron, density, and sonic logs can provide us with porosity values and to estimate water saturation and shale content, resistivity and gamma-ray are employed, respectively. Furthermore, well logging operations are relatively similar in both conventional and unconventional reservoirs while factors such as the extent of the free gas saturation, adsorbed gas capacity, and also NFs characteristics (presence, orientation, intensity, etc.) in URS can be recognized from well log analysis.

One other important point in well logging is establishing a meaningful relationship between well log and core data. These two separate sources of data at different measurement scales, should be correlated and calibrated with one another which eventually supports the characterization effort.

2.2.1.3 Well Test Analysis

Well testing refers to evaluation of the pressure behavior of the wells under various reservoir and well conditions also known as pressure transient analysis (PTA) which will provide us with a wealth of information and properties from the reservoir including reservoir permeability, skin factor, initial reservoir pressure, reservoir flow geometry, reservoir boundary flow type, etc.

Well testing in URs is mainly done under either pre-frac and post-frac conditions. Through the PTA of pre-frac tests, items such as permeability and initial pressure of the virgin reservoir area can be acquired while the post-frac data provides us with HFs' dimensional parameters (height and width) and their conductivity. In post-frac analyses, since flow regimes are being affected due to the presence of HFs on the one hand, and slow development of the radius of investigation of the test pressure pulse in the lateral extent of the reservoir due to the low-permeability nature of these reservoirs on the other hand, the time that is required for proper data recording to later estimate reservoir permeability and initial pressure will be lengthy and not practical. Figure 2.6 demonstrates a sample post-frac test logarithmic diagram of the pseudo-pressure drop versus shut-in time of a gas well in a tight reservoir. The build-up time for this well was designed to be two weeks, although because of the pressure gauge instalment in the wellbore, the pressure data was recorded for 240 days [7]. This curve reveals that even after such a long period of time there is no sign of infinite-acting radial flow (IARF) regime (zero slope of the derivative plot) and the data still represents a linear flow due to the flow through HFs (half-unity slope of the derivative plot).

Since radial flow regime is not achieved in a short period of time and field operators are not inclined to extend the time of the well tests, it is necessary to use pre-frac tests to characterize unstimulated (virgin) areas of the reservoir. However, despite the adequate response time of these tests to determine the permeability and initial pressure of the reservoir, these analyses have the problem of being limited in scale of the study area. Pre-frac tests are divided into open- and cased-hole tests which are classified in Table 2.5 following [7].

In pre-frac tests, the use of mini-frac and diagnostic fracture injection tests (DFIT) in shale and tight gas reservoirs is more common [34]. These two tests are based on creating a fracture in a small-scale section at the interface of the well and the reservoir, measuring the fracture closure pressure, and finally measuring the reservoir pressure and matrix permeability. The fracture closure pressure for vertical wells is equal to the minimum horizontal in-situ stress. Considering Fig. 2.7, which exhibits pressure

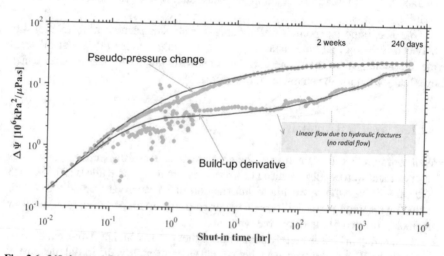

Fig. 2.6 240-day post-frac build-up test plot of a tight gas well and not reaching IARF regime (Modified from [7])

Table 2.5 Different types of open-hole and cased-hole pre-frac tests

Open-/cased-hole test	Type of the test
Open-hole	– Drill stem test (DST) – Mini DST – Wireline formation test (WFT) – Formation rate analyzer (FRA) – Modular dynamic test (MDT) – Repeat formation test (RFT)
Cased-hole	– Slug and impulse tests – Perforation inflow diagnostic (PID) – Perforation inflow test analysis (PITA) – Closed chamber tests – Flow-rate tester (FRT) – Mini-frac tests – Diagnostic fracture injection tests (DFIT)

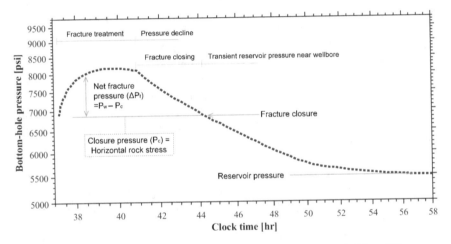

Fig. 2.7 Pressure change versus time plot of a sample mini-frac test (Modified from [7])

variations with the mini-frac test timeframe closure pressure is defined as the pressure that can close the fracture, and can be determined from the fall-off region of the mini-frac test. It should be noted that to keep fracs open, their internal pressure must be kept higher than the closure pressure. This indicates the importance of determining this parameter for HFs' monitoring goals. Also, the important point in mini-frac tests is the possibility of estimating the permeability of the reservoir after reducing the pressure from closure pressure.

More detailed study of PTA methods along with their exact mathematical modeling for well test interpretations is outlined in Chap. 3.

2.2.1.4 Production Analysis

Production analysis, traditionally called decline curve analysis (DCA), and nowadays covers new methods of rate transient analysis (RTA) using the production history of the reservoir. Using empirical DCA models, these methods can determine important parameters for better reservoir characterization such as the production mechanism, the volume of hydrocarbons in-place, and the estimated ultimate recovery (EUR). Also, advanced RTA methods, relying on the analytical/numerical solution of the fluid flow equation in the reservoir, are used to estimate other reservoir parameters as permeability, skin factor, hydraulic fracture height and width, flow regimes, and reservoir model (i.e., being homogenous, naturally fractured).

In general, production analysis methods in URs can be categorized as: (1) straight line analysis, (2) type curve analysis, (3) history matching, and (4) empirical method. Straight line methods are based on analytical solutions, detection and analysis of flow regimes, and determination of HFs and reservoir properties. Type curves are based on matching production data (pressure and flow rate) with pre-defined dimensionless curves (employing an analytical solution of the flow equation) and finally determining the reservoir model, flow regimes, and HFs and reservoir properties. History matching methods, by taking advantage of analytical or numerical solution of the flow equation and consequently flow simulation in a porous medium, are based on matching the production results of the simulated model with the production history from well/wells and thus achieving similar objectives as type curve matching method. Finally, empirical methods based on [2] studies also predict production rates, the volume of hydrocarbon in-place, EUR, and in sometimes flow regime by matching production data with Arps curves.

RTA in URs requires special attention to factors such as; extremely low-permeability of the reservoir and prolonged access to data for reservoir properties calculation, being naturally fractured and the dual or triple porosity behavior of the reservoir, presence of multiple transverse HFs with horizontal well, severe heterogeneities in the reservoir such as multi-layered reservoirs concept and horizontal heterogeneities, the stress dependence of the matrix and NF permeability due to high compressibility, gas absorption, non-Darcy mechanisms of gas flow such as the slip flow and molecular diffusion [7]. It should be noted that following the interpretation of RTA results and determination of HFs and reservoir properties, results should be supported by other sources of data such as micro-seismic, well log and core analyses, as well as production and tracer logs.

Figure 2.8 displays an example of the results of a shale oil well RTA in Bakken field. The left-side plot illustrates the type curve analysis, and the right-side one shows a straight-line analysis plot of linear flow (early and later linear flows) and boundary dominated flow (BDF). These analyses were performed for flow regime identification of the studied well. It should be noticed that if the flow regime is not completely affected by the BDF, estimate for the drainage radius will be made underestimated, therefore, reservoir engineers should be more cautious to avoid similar errors in the estimation of other properties [7].

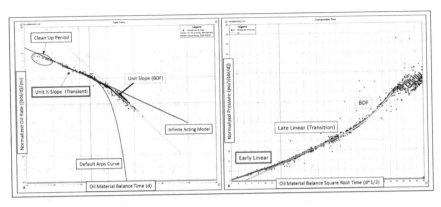

Fig. 2.8 Type curve and straight-line RTA plots for a shale oil well in Bakken field [4]

In the above RTA plots, which are from a shale oil reservoir, normalized pressure data are plotted versus square root of time for straight line analysis (similar to PTA's), and normalized oil production rate data are plotted versus material balance time for type curve analysis (rate decline analysis). It should be noted that a detailed review of production analysis methods including DCA and RTA along with their exact mathematical modeling is presented in Chap. 4.

2.2.2 Hydraulic Fracture Modeling

Comprehensive study of HF models in unconventional formations is not possible by merely studying RTA/PTA methods since they often use simple assumptions to model HFs that might not truly represent the reservoir. Patterns depicted in Fig. 2.9 are examples of different HF models depending on formation ductility/brittleness— simple model, complex model, very complex model, and "tree" fracture network.

Low-complexity fractures mostly can be characterized by RTA/PTA methods, but following Table 2.4, more accurate identification and description of complex fractures

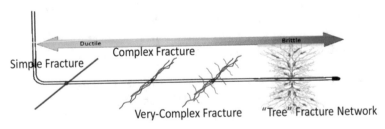

Fig. 2.9 Schematic of different hydraulic fracture models depending on formation ductility/brittleness [36]

are performed by comparing data from other methods, such as micro-seismic and 4D-seismic analyses.

It is worth noting that, [7] explained, modeling HFs in conventional reservoirs, which encompasses recording the pressure during the fracturing job, minimum stress profile, mechanical properties of the rock, fluid properties, injection rate, amount of proppant, and information about the slurry of the fracturing operation, to determine the length and conductivity of the fracture, are not practical in URs due to their operational complexities. Also, it has been outlined that since the results of seismic analysis are not able to determine dynamic parameters such as fracture conductivity (which is dependent on fluid flow), a combination of RTA/PTA and micro-seismic methods would be necessity. In addition, integration of RTA/PTA methods with geomechanical simulation can be used to predict how the fractures open and propagate during the fracturing operation [6].

2.2.3 Reservoir Modeling

In reservoir engineering of URs in particuler, when reservoir and HF characterization is completed using core and well log, or PTA/RTA analyses, reservoir modeling and simulation should be done next for better field development and monitoring. By creating a basic model of the reservoir, which can be deterministic or stochastic [23], future studies on the reservoir will become more reliable, while the basic model can be updated during the life of the reservoir to guarantee maximum accuracy. In the following section, based on Clarkson et al. (2011) [7] and also our own previous studies, some common phenomenon in the modeling and simulation of URs are explained as follows.

1. **Local grid refinement (LGR)**: To enhance precision of the computations in URs modeling and simulation, numerical grid-blocks around the multi-stage fractured horizontal wells (MFHW), as well as time-steps, should be created smaller than regular grid-blocks far away from the MFHW. A sample for LGR technique used around a horizontal well with 5 number of HFs is shown Fig. 2.10.

2. **Description of HFs**: To define HFs in numerical simulation models, LGR itself is a popular technique because of its ease of use [26, 27, 31]. In fact, by applying LGR and considering a HF composed of narrow-width grid-blocks, an HF can be defined in a UR model explicitly by considering optional length, height, and width size.

3. **Importance of RTA**: It is necessary for the early phases of the URs development to use RTA for determination of parameters such as fracs length and conductivity, and postponing complex applications of reservoir simulation to later stages of the development by incorporating more complex methods.

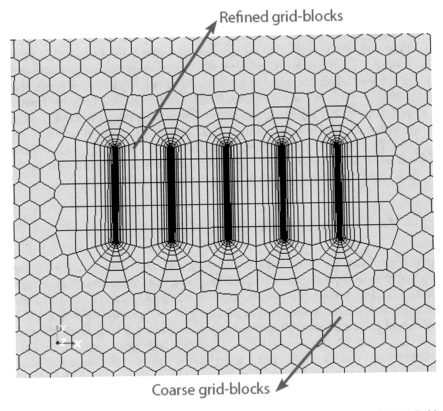

Fig. 2.10 A sample model for UR with a MFHW with 5 number of fracs built in KAPPA Rubis reservoir simulator

4. **Special surveillance data for UGR HM**: UGRs, in addition to the common static (fluid properties, porosity, formation thickness, etc.) and dynamic (bottom-hole pressure and flow rate) data, require more surveillance data for model detection (due to various flow mechanisms) and HM with utmost precision. Micro-seismic data are mainly used to determine the geometry of HFs, and production/tracer logs are used for allocating production rates to different reservoir sections and also for inter-well observations with UGRs' monitoring purposes. As a matter of fact, complexity of this type of reservoirs necessitates more monitoring data to describe and model reservoir behavior.

5. **Handling for uncertainties**: Due to inherent heterogeneity of UR rocks and the consequent uncertainty in static (volumetric) reservoir properties that are obtained from core and well log analyses, a plethora of volumetric data to be used in URs' modeling should be collected. Core and well log data correlation as well as stochastic geostatistical methods are commonly used to take into account static data in URs' modeling. In such a way, properties of certain locations in the

reservoir are estimated using the core-log correlation and geostatistical methods will help us to interpolate between them.

6. **Flowback fluid production simulation**: To create MFHWs in URs, a large volume of fluid is pumped into the reservoir during the fracturing operation. One of the most important issues in these reservoirs is flow simulation (preferably with taking into account for geomechanical effects) to determine the amount of fracturing fluid injection, also flowback water and oil/gas to identify water retention mechanisms. This is important to model gas phase effective permeability changes vs. water imbibition. Proper reservoir modeling and simulation accompanied by reliable data for relative permeability of present fluid phases will reveal the flowback fluids production rate versus time.

2.2.4 Developmental and Economical Planning

After constructing the base model and advancing in simulation with the existing production scenario, other scenarios should be performed and the results of the simulation prediction with each scenario should be examined. Financial considerations for each scenario must be estimated so ultimately the best scenario (lowest cost with the highest amount of revenue) among all possible production scenarios is selected.

As it is argued by Clarkson et al. (2011) [7], some operational parameters related to URs such as frac length and well spacing have evident contribution in final profitability of unconventional fields production. As it was observed by Voneiff and Gatens (1993) [35], although longer frac length is positively correlated to reservoir fluid production rate, further costs due to HFs treatments such as sand production may reduce the final revenue from the fractured wells. In addition, increasing the number of MHFWs in a field (enhancing the fracs density in the total field) has a decreasing effect—known as offset frac hits—on the production of the wells compared to production of each one in the absence of other wells. Frac hits may cause sudden pressure spikes in recorded data, water cut increase, and also oil/gas production decrease [15]. Considering such operational-economic effects in development of URs, an optimization procedure should be carried out to reveal parameters/considerations to minimize costs and optimize the revenue.

References

1. Ahmed U, Meehand D (2016) Unconventional oil and gas resources. CRC Press, Boca Raton, Florida, Explotation and Development
2. Arps JJ (1945) Analysis of decline curves. Trans AIME 160:228–247
3. Bae JS, Bhatia SK (2006) High-pressure adsorption of methane and carbon dioxide on coal. Energy Fuels. https://doi.org/10.1021/ef060318y
4. Brown CL (2018) Investigating the impact of offset fracture hits using rate transient analysis in the Bakken and three forks formation, divide county. University of North Dakota, North Dakota

5. Brunauer S, Emmett PH, Teller E (1938) Adsorption of gases in multimolecular layers. J Am Chem Soc 60:309–312
6. Cipolla CL, Williams MJ, Weng X, et al (2010) Hydraulic fracture monitoring to reservoir simulation: maximizing value. In: Second EAGE middle east tight gas reservoirs workshop. European Association of Geoscientists & Engineers
7. Clarkson CR, Jensen JL, Blasingame TA (2011) Reservoir Engineering for Unconventional Gas Reservoirs: What Do We Have to Consider? North Am. Unconv. Gas Conf. Exhib
8. Cui X, Bustin AMM, Bustin RM (2009) Measurements of gas permeability and diffusivity of tight reservoir rocks: different approaches and their applications. Geofluids 9:208–223
9. Curtis JB (2002) Fractured shale-gas systems. Am Assoc Pet Geol Bull 86:1921–1938
10. Dahim S, Taghavinejad A, Razghandi M et al (2020) Pressure and rate transient modeling of multi fractured horizontal wells in shale gas condensate reservoirs. J Pet Sci Eng 185:106566. https://doi.org/10.1016/j.petrol.2019.106566
11. Dong JJ, Hsu JY, Wu WJ et al (2010) Stress dependence of the permeability and porosity of sandstone and shale from TCDP Hole-A. Int J Rock Mech Min Sci Geomech Abstr 47:1141–1157
12. Dubinin MM, Astakhov VA (1971) Development of the concepts of volume filling of micro-pores in the adsorption of gases and vapors by microporous adsorbents. Bull Acad Sci USSR Div Chem Sci. https://doi.org/10.1007/bf00849308
13. Javadpour F (2009) Nanopores and apparent permeability of gas flow in mudrocks (shales and siltstone). J Can Pet Technol 48:16–21
14. Klinkenberg LJ (1941) The permeability of porous media to liquids and gases. Drill Prod Pract
15. Kurtoglu B, Salman A (2015) How to utilize hydraulic fracture interference to improve unconventional development. In: Society of petroleum engineers—Abu Dhabi international petroleum exhibition and conference, ADIPEC 2015
16. Langmuir L (1918) The adsorption of gases on plane surfaces of glass, mica and platinum. J Am Chem Soc 40:1361–1403
17. Lee KS, Kim TH (2016) Integrative understanding of shale gas reservoirs. Springer
18. Li J, Lu S, Cai J, et al (2018) Adsorbed and free oil in Lacustrine Nanoporous Shale: a theoretical model and a case study. Energy Fuels 32:12247–12258. https://doi.org/10.1021/acs.energy fuels.8b02953
19. Liu J, Zhao Y, Yang Y et al (2020) Multicomponent shale oil flow in real kerogen structures via molecular dynamic simulation. Energies 13:3815. https://doi.org/10.3390/en13153815
20. Palmer I, Mansoori J (1998) How permeability depends on stress and pore pressure in coalbeds: a new model. SPE Reserv Eval Eng 1:539–544
21 Raghavan R, Chin LY (2004) Productivity changes in reservoirs with stress-dependent permeability. SPE Reserv Eval Eng 7:308–315
22. Rapp BE (2017) Fluids. In: Rapp BE (ed) Microfluidics: modelling. Elsevier, Mechanics and Mathematics, pp 243–263
23. Rushing JA, Newsham KE (2001) An integrated work-flow model to characterize uncon-ventional gas resources: Part II-formation evaluation and reservoir modeling. In: SPE annual technical conference and exhibition. Society of Petroleum Engineers
24. Seidle JR, Huitt LG (1995) Experimental measurement of coal matrix shrinkage due to gas desorption and implications for cleat permeability increases. Int Meet Pet Eng
25. Shi JQ, Durucan S (2004) Drawdown induced changes in permeability of coalbeds: a new interpretation of the reservoir response to primary recovery. Transp Porous Media. https://doi.org/10.1023/B:TIPM.0000018398.19928.5a
26. Shirbazo A, Fahimpour J, Aminshahidy B (2020) A new approach to finding effective parame-ters controlling the performance of multi-stage fractured horizontal wells in low-permeability heavy-oil reservoirs using RSM technique. J Pet Explor Prod Technol 10:3569–3586. https://doi.org/10.1007/s13202-020-00931-3
27. Shirbazo A, Taghavinejad A, Bagheri S (2021) CO_2 Capture and Storage Performance Simu-lation in Depleted Shale Gas Reservoirs as Sustainable Carbon Resources. J Constr Mater. https://doi.org/10.36756/JCM.si1.3

28. Sondergeld CH, Newsham KE, Comisky JT, et al (2010) Petrophysical considerations in evaluating and producing shale gas resources. In: SPE unconventional gas conference. Society of Petroleum Engineers

29. Song B (2010) Pressure transient analysis and production analysis for New Albany shale gas wells. Texas A&M University

30. Sunden B, Fu J (2017) Low-density heat transfer: rarefied gas heat transfer. In: Sunden B, Fu J (eds) Heat transfer in aerospace applications. Academic Press, pp 45–70

31. Taghavinejad A, Shafeie S, Shirbazo A (2021) Analysis of wastewater disposal in depleted tight gas reservoirs: a sustainable resources approach. J Constr Mater I Spec Issue. https://doi.org/10.36756/JCM.si1.4

32. Taghavinejad A, Sharifi M (2021) Investigation of rock properties distribution effect on pressure transient analysis of naturally fractured reservoirs. J Pet Sci Eng 204:108714. https://doi.org/10.1016/j.petrol.2021.108714

33. Taghavinejad A, Sharifi M, Heidaryan E et al (2020) Flow modeling in shale gas reservoirs: a comprehensive review. J Nat Gas Sci Eng 83:103535. https://doi.org/10.1016/j.jngse.2020.103535

34. Thompson D, Rispler KA, Stadnyk SM, et al (2009) Operators evaluate various stimulation methods for multizone stimulation of horizontals in North East British Columbia. In: SPE hydraulic fracturing technology conference. Society of Petroleum Engineers

35. Voneiff GW, Gatens JM (1993) Benefits of applying technology to Devonian Shale wells. In: Proceedings—SPE Eastern regional conference and exhibition. Society of Petroleum Engineers (SPE), pp 45–56

36. Wang D, Ge H, Wang X et al (2015) A novel experimental approach for fracability evaluation in tight-gas reservoirs. J Nat Gas Sci Eng 23:239–249. https://doi.org/10.1016/j.jngse.2015.01.039

37. Wang S, Feng Q, Javadpour F et al (2015) Oil adsorption in shale nanopores and its effect on recoverable oil-in-place. Int J Coal Geol 147–148:9–24. https://doi.org/10.1016/j.coal.2015.06.002

38. Warren JE, Root PJ (1963) The behavior of naturally fractured reservoirs. SPE J 3:245–255. https://doi.org/10.2118/426-PA

39. Yang Y, Liu J, Yao J et al (2020) Adsorption behaviors of shale oil in kerogen slit by molecular simulation. Chem Eng J 387:124054. https://doi.org/10.1016/j.cej.2020.124054

40. Zhao Y, Zhang L, Zhao J et al (2013) "Triple porosity" modeling of transient well test and rate decline analysis for multi-fractured horizontal well in shale gas reservoirs. J Pet Sci Eng 110:253–262

Chapter 3
Pressure Transient Analysis

3.1 Pre-Frac Well Test

Pre-frac tests are short-term well-testing techniques that are used in unconventional reservoirs (URs) to obtain reservoir properties and a means of data collection via pressure transient analysis (PTA). These PTA techniques have much shorter test duration compared to build up and long-term draw-down tests which are mostly run for at least several weeks. As an example, the mini drill stem testing (mini-DST) consists of only a production period of 20 to 30 min and then 1-h of build-up test [12]. As a matter of fact, pre-frac tests by having shorter test duration, especially compared to conventional methods that are commonly used for URs, are considered as one of the best approaches to determine intact (unstimulated) properties of the reservoir. However, their radius of investigation is limited, and the collected data is relatively restricted to the area around the test location (not the entire reservoir). In the following sections, two important pre-frac tests; mini-DST and mini-frac in low-permeability formations will be explained in detail from technical point of view. It should be noted that most of these tests are carried out in tight/shale oil reservoirs, with corresponding mathematical equations. However, following similar approach, formulation can also be driven for unconventional gas reservoirs (UGRs) too.

3.1.1 Mini-DST

A typical DST is based on the production and/or pressure build-up after well shut-in period, and subsequently calculating reservoir properties—permeability and skin factor. Though, because of low-permeability of URs, radius of investigation of the mini-DSTs should be designed smaller than a typical DST to shorten the test duration. Mini-DST is based on producing small volumes of fluid from a single perforation into a chamber and then followed by the build-up analysis. Figure 3.1 shows a

© The Author(s), under exclusive license to Springer Nature Switzerland AG 2022
A. Taghavinejad et al., *Unconventional Reservoirs: Rate and Pressure Transient Analysis Techniques*, SpringerBriefs in Petroleum Geoscience & Engineering, https://doi.org/10.1007/978-3-030-82837-0_3

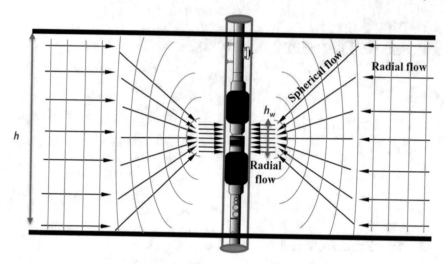

Fig. 3.1 Schematic view of a mini-DST and its occurring flow regimes in an open-hole completion [8]

schematic view of Mini-DST and the flow geometries emerged during this type of PTA technique. As it is inferred from this figure, initial radial flow, spherical flow, and—if the test continues longer—secondary radial flow regimes are expected to be seen during a mini-DST. Required tools for implementing a mini-DST include fluid inlet chamber, constant flowrate pump, and pressure gauge.

Mathematical model for the pressure change in spherical flow regimes in oil reservoirs was developed by Joseph and Koederitz (1985) [7] as Eq. 3.1 for drawdown and Eq. 3.5 for build-up tests.

$$\Delta P = \frac{70.6q\mu B}{k_{sp}r_{sw}}(1+S) - \frac{2453q\mu B\sqrt{\mu c_t \varphi}}{k_{sp}^{3/2}}\frac{1}{\sqrt{t}} \qquad (3.1)$$

where q is the flow rate (STB/day), μ is fluid viscosity (cp), B is fluid volume coefficient (bbl/STB), k_{sp} is spherical permeability (md), r_{sw} is pseudo-radius of spherical wellhead (ft), φ is formation porosity (ratio), and t is test time (h), all in field unit. Also, ΔP in this equation is the difference between the initial pressure and the bottom-hole pressure. Moreover, r_{sw} can be calculated using the following equation.

$$r_{sw} = \frac{0.5h_w}{\ln\left(\frac{h_w}{r_w}\right)} \qquad (3.2)$$

where h_w refers to the interval in the formation that is restricted between two packers during in the mini-DST.

Finally, for a draw-down mini-DST, by plotting ΔP versus $t^{-0.5}$, the slope of the line will be:

$$m = \frac{2453q\mu B\sqrt{\mu c_t \varphi}}{k_{sp}^{3/2}} \tag{3.3}$$

If all parameters are known in Eq. 3.3, the spherical permeability k_{sp} can be calculated. Moreover, the following relationship is established between permeability in x, y, and z directions, or in other words, between the radial (k_r) and vertical (k_v) permeabilities with the spherical permeability.

$$k_{sp} = \sqrt[3]{k_x k_y k_z} = \sqrt[3]{k_r^2 k_v} \tag{3.4}$$

By shutting in the well to stop the flow and starting a build-up test in a mini-DST, PTA formula can be written as Eq. 3.5.

$$P_{ws}(\Delta t) - P_{wf}(\Delta t = 0) = \frac{2453q\mu B\sqrt{\mu c_t \varphi}}{k_{sp}^{3/2}}\left(\frac{1}{\sqrt{t_p + \Delta t}} - \frac{1}{\sqrt{\Delta t}}\right) \tag{3.5}$$

where P_{ws} is well shut-in pressure, P_{wf} is well-flowing pressure, t_p is production period time, and Δt is the build-up time after production period.

According to Fig. 3.1 and as explained by Kurtoglu (2013) [8], since flow will farther propagates in the reservoir, flow lines will change to radial flow patterns. Then, by calculating the permeability in the form of conventional well testing, radial permeability is obtained enabling us to estimate vertical permeability using Eq. 3.4. Infinite acting radial flow (IARF) PTA formula obtained in mini-DST is defined by Kurtoglu as follows:

$$\Delta P = P_i - P_{wf}(t) = \frac{162.6q\mu B}{kh}\left(\log(t) + \log\left(\frac{k}{\mu c_t \varphi r_w^2}\right) - 3.228 + 0.8686S\right) \tag{3.6}$$

where h is the total thickness of the formation. It is evident that by plotting ΔP versus $\log(t)$, the slope of the resulting line will be Eq. 3.7 providing the radial permeability (k_r).

$$m = \frac{162.6q\mu B}{k_r h} \tag{3.7}$$

In addition, straight-line analysis of initial radial flow—prior to the spherical flow—is obtained by plotting ΔP versus $\log(t)$ and calculating the slope as:

$$m = \frac{162.6q\mu B}{k_r h_w} \tag{3.8}$$

where h_w is the production interval of mini-DST apparatus (ft) as illustrated in Fig. 3.1.

Considering radial permeability concept, the following equation is established between permeability in x (k_x) and y (k_y) directions, and radial permeability.

$$k_r = \sqrt{k_x k_y} \tag{3.9}$$

It should be noted that since there is a relationship between pressure variation and the second root of time $(t^{-0.5})$ in spherical flow, this particular flow regime in all logarithmic and type curve plots will be a line with the slope $= -0.5$.

3.1.2 Mini-Frac Test

Mini-frac test is a type of injection and then fall-off test (injection fall-off test) for both vertical and horizontal wells (HWs), which is used to estimate the economic feasibility of drilling in the tested formation by measuring the closure pressure parameters (P_c), leak-off coefficient (C_L), and other reservoir properties such as effective permeability and reservoir pressure [8].

The plot of pressure and flowrate versus time for a sample mini-frac test is depicted in Fig. 3.2. It is understood from this figure that this test is initially associated with an increase in injection flowrate as well as an increase in bottom-hole pressure (a). Then, when the flow pressure reaches the formation breakdown pressure T, a fracture is created in the formation and by stabilizing the flow rate, the amount of bottom-hole pressure experiences a slow decrease, and fracture propagation continues in this period (b). The fracture is propagated and the flow is fallen off to stop the fracture development, but not immediately and for a while the fracture aperture is increased which will cause a sharper decline in the pressure (c). This is followed by the pressure reaching the fracture closure pressure (P_c) causing the pressure decline to continue for a while due to the falling off the injection flow (d). Finally, the bottom-hole pressure is stabilized and properties such as permeability and formation pressure can be extracted from this phase (e).

Closure pressure, P_c, is the pressure that closes the fracture after its pressure is reduced and is equivalent to the minimum horizontal stress in vertical wells. By knowing its magnitude, we can keep the internal pressure of the fracture higher than the closure pressure to prevent fracture from closing. According to Fig. 3.2, the amount of P_c can be calculated after the injection fall-off—at the onset of pressure decline which is equivalent to the interface of regions b and c in Fig. 3.2. A closer look at the mini-frac test pressure curve reveals that the amount of P_c is not exactly correspondent to the flow fall-off pressure. In fact, fractures that are created will

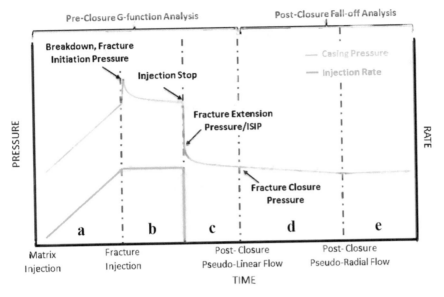

Fig. 3.2 Pressure and flowrate plot of a sample mini-frac test [8]

continue to propagate after the flow is stopped, and bottom-hole pressure has reached the closure pressure for some time followed by flow fall-off. P_c is determined as the pressure equivalent to the inflection point of the pressure curve following the injection flow fall-off.

It was shown in Fig. 3.2 that the magnitude of formation breakdown pressure can be obtained from mini-frac test. Moreover, formation breakdown pressure, can also be calculated using Eq. 3.10 for vertical wells and Eqs. 3.11 and 3.12 for HWs as well [12].

$$P_{bd} = 3\sigma_h - \sigma_H - P + T \qquad (3.10)$$

$$P_{bd} = 3\sigma_h - \sigma_v - P + T \qquad (3.11)$$

$$P_{bd} = 3\sigma_H - \sigma_v - P + T \qquad (3.12)$$

In above equations; σ_h is minimum horizontal stress, σ_H is maximum horizontal stress, σ_v is vertical stress, P is formation pressure, and T is tensile strength of the formation. Here, Eq. 3.11 is related to HWs drilled in the direction of σ_H, and Eq. 3.12 is representing HWs drilled in the direction of σ_h. As it is explained by Rezaee (2015) [12], if a HW is drilled in the direction of σ_h and also σ_H is much larger than σ_h, the formation breakdown pressure will become very large, which is common in URs.

The initial step in the interpretation of a mini-frac test data, would be the analysis of the closure pressure (P_c) after flow fall-off. Additionally, pore pressure (P) estimation can be done when the bottom-hole pressure is fully stabilized after the flow is stopped by plotting bottom-hole pressure versus time. Ultimately to calculate pore pressure, permeability relationships developed by Economides and Nolte (1987) [4] can be utilized as follows.

$$\Delta P = \frac{\pi C_L \sqrt{t_p}}{2c_f} G(\Delta t_D)$$ (3.13)

where:

$$\Delta P = P_w(\Delta t_D = 0) - P_w(\Delta t_D)$$ (3.14)

$$G(\Delta t_D) = \frac{16}{3\pi}\left[(1 + \Delta t_D)^{\frac{3}{2}} - \Delta t_D^{\frac{3}{2}} - 1\right]$$ (3.15)

$$\Delta t_D = \frac{t - t_p}{t_p} = \frac{\Delta t}{t_p}$$ (3.16)

where, t_p is the pumping time of the fluid and t is the elapsed time since the start of the formation breakdown and both are in minutes (min). Also, C_L is leak-off coefficient (ft/$\sqrt{min}$) and c_f is the fracture compliance (ft/psi). It should be noted that C_L has a positive correlation with the fluid leak-off to the formation.

By plotting ΔP versus G-function ($G(\Delta t_D)$), a straight-line is attained with slope m as Eq. 3.17.

$$m = \frac{\pi C_L \sqrt{t_p}}{2c_f}$$ (3.17)

And finally, c_f can be calculated using Eq. 3.18.

$$c_f = \frac{\pi(1 - v^2)}{2E} \frac{q\sqrt{t_p}}{2\pi h_f C_L}$$ (3.18)

where, v is Poisson's ratio, E is Young's modulus, q is injection flow rate (ft/min) and h_f is fracture height (ft). By determining C_L, formation matrix permeability (k_m) can be found using the following relationship for C_L.

$$C_L = 0.00118\Delta P_c\sqrt{\frac{k_m c_t \varphi}{\mu}}$$ (3.19)

In the above equations, all parameters are in field unit and also ΔP_c is defined as the difference between the closure pressure of the fracture and the pore pressure of the formation ($\Delta P_c = P_c - P$).

Furthermore, the average fracture width during fluid injection is obtained from Eq. 3.20.

$$\overline{w}_f = c_f(P_{hf} - P_c) \tag{3.20}$$

where, P_{hf} denotes the fracture propagation pressure in the test.

3.2 Post-Frac Well Test

PTA in hydraulically fractured wells in URs is usually conducted using long-term build-up tests data. In this regard, low permeability of the reservoirs causes a long build-up duration. This well cause the propagation of pressure pulse in the zone of investigation in the reservoir to increase. Ultimately, this makes such tests to be overall longer in period in unconventional reservoirs compared to conventional ones. Hence, post-frac well tests are replaced with RTA techniques because long period of well shut-in is not economically justified.

Pressure analyses, depending on the type of hydraulic fractures (HFs) that are created in the well and reservoir system itself, can be used for both finite- and infinite-conductivity fractures as bilinear and linear flow regimes, respectively, that is discussed below.

3.2.1 Bilinear Flow Analysis

The bilinear flow regime usually follows the flow of a fluid from a well with a finite-conductivity hydraulic fracture (FCHF). However, this flow pattern can be seen in early-time transient flow from HWs [1] and the transient flow of wells located in low-permeability dual-porosity reservoirs [8]. Different cases of bilinear flow PTA in URs are explained later.

3.2.1.1 Bilinear Flow Analysis in FCHF

If the production is taking place from wells with FCHFs, flow patterns appear as Fig. 3.3 in the HF and its surroundings. At the beginning of the production flow, the linear flow of the fracture is formed according to Fig. 3.3a and then, as the pressure pulse of the bilinear flow continues to propagate, it is formed as Fig. 3.3b in the reservoir extent. At the end of the bilinear flow regime, the linear flow pattern of the formation appears towards the HF as Fig. 3.3c, and ultimately, with increase in the propagation of pressure-drop pulse in the well and reservoir system, pseudo-radial flow regime according to Fig. 3.3d is seen around the HF.

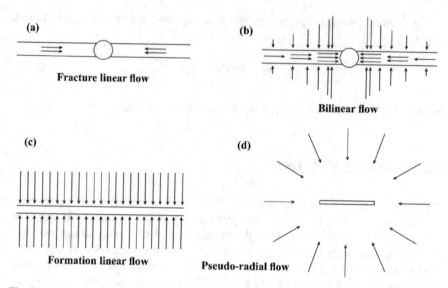

Fig. 3.3 Schematic view of the fracture linear (**a**), bilinear (**b**), formation linear (**c**), and pseudo-radial (**d**) flows in a HW with FCHF (modified from [14])

As [12] illustrated, the total pressure drop of the bilinear flow (before the linear flow of the formation) can be calculated by the following equation:

$$\frac{\Delta P(t)}{q_o B_o + q_w B_w + q_g B_g} = \frac{44.102}{h n_{hf} \lambda_t \sqrt{w_f k_{hf}}} \left(\frac{\lambda_t}{c_t \varphi k_{eff}}\right)^{\frac{1}{4}} t^{\frac{1}{4}} + \frac{141.2 S_{hf}^{well}}{h n_{hf} k_{eff} \lambda_t} \qquad (3.21)$$

where in this equation; q_o, q_w, and q_g are production flowrate of oil, water and gas, respectively, B_o, B_w, and B_g are formation volume factors of oil, water, and gas, respectively, h is formation thickness, n_{hf} is the number of HFs, w_f is HF width, k_{hf} is HF permeability, c_t is total compressibility of the formation, φ is formation porosity, k is formation permeability, S_{hf}^{well} is skin factor of well-HF connection, and λ_t are summation of inverse of oil, water, and gas viscosities ($1/\mu_o + 1/\mu_w + 1/\mu_g$), and all parameters are in field unit.

Equation 3.21 expresses the pressure relationship for a draw-down test for wells with FCHF. Considering the build-up test, Eq. 3.22 can be used in similar wells and reservoir systems.

$$\frac{P_{ws}(\Delta t) - P_{ws}(\Delta t = 0)}{q_o B_o + q_w B_w + q_g B_g} = \frac{44.102}{h n_{hf} \lambda_t \sqrt{w_f k_{hf}}} \left(\frac{\lambda_t}{c_t \varphi k_{eff}}\right)^{\frac{1}{4}} \left[(t_p + \Delta t)^{\frac{1}{4}} - (\Delta t)^{\frac{1}{4}}\right]$$

$$(3.22)$$

In this equation, P_{ws} is well shut-in pressure, Δt is the test time after well shut-in, and t_p is the production time from the well before well shut-in.

Table 3.1 Straight-line analyses of bilinear flow well test data in wells with FCHF

Straight-line analysis plot	Test type
$\dfrac{\Delta P(t)}{q_o B_o + q_w B_w + q_g B_g}$ versus $t^{\frac{1}{4}}$	Draw-down
$\dfrac{P_{ws}(\Delta t) - P_{ws}(\Delta t=0)}{q_o B_o + q_w B_w + q_g B_g}$ versus $\left[(t_p + \Delta t)^{\frac{1}{4}} - (\Delta t)^{\frac{1}{4}} \right]$	Build-up
Slope of the lines: $m = \dfrac{44.102}{h n_{hf} \lambda_t \sqrt{w_f k_{hf}}} \left(\dfrac{\lambda_f}{c_t \varphi k_{eff}} \right)^{\frac{1}{4}}$	

Since most URs are naturally fractured, the following relationships can be used to estimate the permeability (k_{eff}) and reservoir flow capacity ($c_t \varphi$) of unconventional plays using Eq. 3.23 and 3.24 [12].

$$c_t \varphi = (c_t \varphi)_f + (c_t \varphi)_m \tag{3.23}$$

$$k_{eff} = k_f \varphi_f + k_m \varphi_m \tag{3.24}$$

where subscripts "f" and "m" are related to the fracture and matrix media of the UR rock, respectively.

It should be noted that in the draw-down and build-up tests of above PTA, the analysis of the straight-line by plotting the pressure/production versus the fourth-root of time as summarized in Table 3.1 can provide us with the magnitude of $w_f k_{hf}$ (frac conductivity), if all other parameters are known.

3.2.1.2 Bilinear Flow Analysis in HW with ICHF in a Dual-Porosity Reservoir

In addition to the analysis of the flow regime resulting from the FCHF using the flow equations of a homogeneous reservoir, pressure analysis of bilinear flow regime in a dual-porosity media has also been developed by Torcuk et al. (2013) [15] which follows the production via ICHFs. This type of flow regime in the formation occurs because linear flow by ICHFs and linear flow of NFs interact, which is schematically shown in Fig. 3.4 [8].

Following the study by Rezaee (2015) [12], Eq. 3.25 expresses the pressure-drop relationship due to the production from ICHF in a dual-porosity reservoir with bilinear flow pattern.

$$\frac{\Delta P(t)}{q_o B_o + q_w B_w + q_g B_g} =$$

$$\frac{45.103}{h n_{hf} L_f \lambda_t \sqrt{k_{eff}}} \left[\left(\frac{1}{1-\omega_f} \right)^{\frac{1}{2}} \frac{1}{\left(\frac{\alpha}{4} k_m (c_t \varphi \lambda_t^{-1})_{f+m} \right)^{\frac{1}{4}}} \right] t^{\frac{1}{4}} + \frac{141.2 S_{hf}^{well}}{h n_{hf} k_{eff} \lambda_t} \tag{3.25}$$

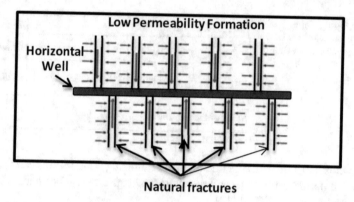

Fig. 3.4 Schematic view of bilinear flow in a HW with ICHF in a dual-porosity reservoir [8]

In this equation, L_f is the frac half-length and α is shape factor of the fractured media.

Equation of pressure changes in this flow geometry for such well and reservoir system for a build-up test is in the form of Eq. 3.26.

$$\frac{P_{ws}(\Delta t) - P_{ws}(\Delta t = 0)}{q_o B_o + q_w B_w + q_g B_g} =$$

$$\frac{45.103}{h n_{hf} L_f \lambda_t \sqrt{k_{eff}}} \left[\left(\frac{1}{1-\omega_f} \right)^{\frac{1}{2}} \frac{1}{\left(\frac{\alpha}{4} k_m \left(c_t \varphi \lambda_t^{-1} \right)_{f+m} \right)^{\frac{1}{4}}} \right] \left[(t_p + \Delta t)^{\frac{1}{4}} - (\Delta t)^{\frac{1}{4}} \right]$$

(3.26)

Graphical analysis of draw-down and build-up tests in wells with ICHFs in dual-porosity URs (Table 3.1), will enable us to find the slope of the line. Therefore, if all parameters are known, $L_f \sqrt{k_{eff}}$ can be determined.

$$\frac{P_{ws}(\Delta t) - P_{ws}(\Delta t = 0)}{q_o B_o + q_w B_w + q_g B_g} =$$

$$\frac{45.103}{h n_{hf} L_f \lambda_t \sqrt{k_{eff}}} \left[\left(\frac{1}{1-\omega_f} \right)^{\frac{1}{2}} \frac{1}{\left(\frac{\alpha}{4} k_m \left(c_t \varphi \lambda_t^{-1} \right)_{f+m} \right)^{\frac{1}{4}}} \right] \left[(t_p + \Delta t)^{\frac{1}{4}} - (\Delta t)^{\frac{1}{4}} \right]$$

(3.27)

It should be noted that effective permeability (k_{eff}) obtained from this analysis is a combination of the HF, NF, and UR rock matrix permeabilities.

3.2.1.3 Bilinear Flow Analysis in Unstimulated HW in a Dual-Porosity Reservoir

This type of bilinear flow is due to the presence of NFs in the reservoir which are formed as the result of combination of NF-HW linear flow and also matrix-NF linear flow based on Fig. 3.5. Under these conditions, a bilinear flow occurs in a HW without any frac in a dual-porosity reservoir.

Permeability obtained from the PTA of these wells and reservoir systems would be the effective permeability of a system which is basically the average permeability of NF and rock matrix [8]. In the following, based on the work done by Du and Stewart (1992) [3], the relationships between pressures resulting from draw-down and build-up tests are presented in Eqs. 3.28 and 3.29, respectively.

$$\frac{\Delta P(t)}{q_o B_o + q_w B_w + q_g B_g} = \frac{45.103}{h n_{hf} L \lambda_t \sqrt{k_{eff}}} \left[\left(\frac{1}{1-\omega_f} \right)^{\frac{1}{2}} \frac{1}{\left(\frac{\alpha}{4} k_m \left(c_t \varphi \lambda_t^{-1} \right)_{f+m} \right)^{\frac{1}{4}}} \right] t^{\frac{1}{4}}$$

(3.28)

$$\frac{P_{ws}(\Delta t) - P_{ws}(\Delta t = 0)}{q_o B_o + q_w B_w + q_g B_g} =$$

$$\frac{45.103}{h n_{hf} L \lambda_t \sqrt{k_{eff}}} \left[\left(\frac{1}{1-\omega_f} \right)^{\frac{1}{2}} \frac{1}{\left(\frac{\alpha}{4} k_m \left(c_t \varphi \lambda_t^{-1} \right)_{f+m} \right)^{\frac{1}{4}}} \right] \left[(t_p + \Delta t)^{\frac{1}{4}} - (\Delta t)^{\frac{1}{4}} \right]$$

(3.29)

It is noticeable that PTA formulations for bilinear flow in Unstimulated HWs in a dual-porosity reservoir (Eqs. 3.28 and 3.29) are very similar to the pressure of bilinear flow analysis of HWs with ICHF in a dual-porosity reservoir (Eqs. 3.26 and 3.27). The difference is that the flow geometry length parameter in Eqs. 3.26 and

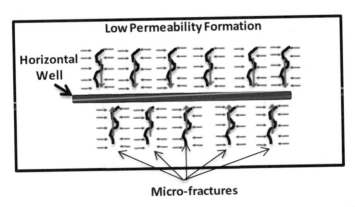

Fig. 3.5 Schematic of bilinear flow in an unstimulated HW in a dual-porosity reservoir [8]

3.27 is the frac half-length (L_f) whereas in the above equations is the HW half-length (L).

By performing straight-line analysis for draw-down and build-up tests according to Table 3.1, the slope of the straight-line will be as follows.

$$m = \frac{45.103}{hn_{hf}L\lambda_t\sqrt{k_{eff}}}\left[\left(\frac{1}{1-\omega_f}\right)^{\frac{1}{2}}\frac{1}{\left(\frac{\alpha}{4}k_m\left(c_t\varphi\lambda_t^{-1}\right)_{f+m}\right)^{\frac{1}{4}}}\right] \tag{3.30}$$

Finally, if all other parameters are known, the effective permeability (k_{eff}) can be calculated from the slope of the straight-line.

Figure 3.6 displays an example of a build-up test with ten days of production and three days of shut-in period in a 9123-foot unstimulated HW in the Bakken Formation. Average oil flowrate in draw-down period is 125 bbl/day. Figure 3.6a shows the pressure history (upper part) and also flowrate history (lower part), and Fig. 3.6b depicts the pressure (upper diagram) and pressure derivative (lower diagram) values versus time.

In the logarithmic plot of Fig. 3.6b, the slope is found 0.25 which is a characteristic of bilinear flow. It can also be seen in Fig. 3.7 that effective permeability of the fracture is estimated by plotting the bottom-hole pressure data versus the fourth root of time.

In this real-world example, two interpretations can be made to determine the effective fracture permeability. As it is outlined by Kurtoglu [8], from the results obtained from this test, displayed in Fig. 3.7, by considering 20% of the well length as the effective length, effective permeability of the fracture is obtained 0.238 md and by considering the entire length of the HW, effective permeability of the fracture is obtained 0.000951 md.

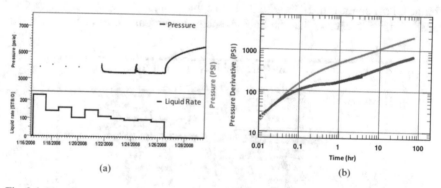

(a)

(b)

Fig. 3.6 Plot of pressure and flowrate history (**a**) and plot of pressure and pressure derivative (**b**) of a build-up test in an unstimulated HW pressure in the Bakken field [8]

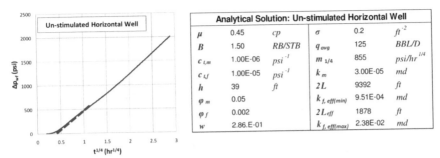

Analytical Solution: Un-stimulated Horizontal Well					
μ	0.45	cp	σ	0.2	ft^{-2}
B	1.50	RB/STB	q_{avg}	125	BBL/D
$c_{t,m}$	1.00E-06	psi^{-1}	$m_{1/4}$	855	$psi/hr^{1/4}$
$c_{t,f}$	1.00E-05	psi^{-1}	k_m	3.00E-05	md
h	39	ft	$2L$	9392	ft
φ_m	0.05		$k_{f,eff(min)}$	9.51E-04	md
φ_f	0.002		$2L_{eff}$	1878	ft
w	2.86.E-01		$k_{f,eff(max)}$	2.38E-02	md

Fig. 3.7 Bilinear flow analysis of unstimulated HW build-up test in the Bakken field [8]

3.2.2 Linear Flow Analysis

Linear flow regime refers to the flow pattern that occurs following the convergence of flow lines in a horizontal plane [9]. The linear flow regime is mostly seen in wells with ICHFs. However, linear flow regime can happen in stimulated and unstimulated horizontal wells in dual-porosity reservoirs too. In the following section, different PTA models for several linear flow applications in URs are explained.

3.2.2.1 Linear Flow Analysis in ICHF

When flow is happening through an ICHF, a linear flow is formed in the well and the reservoir system as shown in Fig. 3.8. This linear flow that takes place from the formation to the HF has a specific pressure analysis, and will provide us with important information such as reservoir permeability and frac length.

To determine the exact amount of pressure changes caused by the production of three-phase fluid (water, oil, and gas) through the ICHF, it is necessary to solve diffusivity equation analytically/numerically in a linear geometry. Hence, diffusivity equation is given for an oil reservoir with water and gas associated phases as Eq. 3.31 and for the gas condensate reservoir as Eq. 3.32 [12].

$$\frac{\partial^2}{\partial L^2}(k\lambda_t P_o) + \left(B_w q_w + B_o q_o + B_g q_g\right) = c_t \varphi \frac{\partial P_o}{\partial t} \qquad (3.31)$$

Fig. 3.8 Schematic of the linear flow of the formation in an ICHF (Modified from [14])

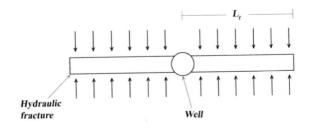

$$\frac{\partial^2}{\partial L^2}(k\lambda_t P_o) + \left(B_w q_w + B_g\left(380\frac{\rho_o}{M_o}q_o + q_g\right)\right) = c_t\varphi\frac{\partial P_o}{\partial t} \qquad (3.32)$$

where; ρ_o is oil density and M_o is molecular weight of oil. By solving oil and gas condensate flow equations in URs producing from wells with ICHFs, pressure response for oil and gas condensate will be Eqs. 3.33 and 3.34, respectively.

$$\frac{\Delta P(t)}{q_o B_o + q_w B_w + q_g B_g} = \frac{4.064\pi\sqrt{t}}{2hn_{hf}L_f\sqrt{k_{eff}\lambda_t}\sqrt{(c_t\varphi)_{f+m}}} + \frac{141.2S_{hf}^{face}}{hn_{hf}k_{eff}\lambda_t} \qquad (3.33)$$

$$\frac{\Delta P(t)}{q_w B_w + \left(380\frac{\rho_o}{M_o}q_o + q_g\right)B_g} = \frac{4.064\pi\sqrt{t}}{2hn_{hf}L_f\sqrt{k_{eff}\lambda_t}\sqrt{(c_t\varphi)_{f+m}}} + \frac{141.2S_{hf}^{face}}{hn_{hf}k_{eff}\lambda_t} \qquad (3.34)$$

where S_{hf}^{face} is skin factor due to the connection of HF and reservoir.

The slope obtained from the straight-line analysis of plotting the left hand-side (LHS) terms of above expressions versus $\sqrt{t}$, effective permeability (k_{eff}) can be calculated if other parameters are already known.

3.2.2.2   Linear Flow Analysis in Unstimulated HW in a Dual-Porosity Reservoir

As mentioned earlier, a linear flow regime in unstimulated HWs in dual-porosity reservoirs can also be observed. This type of linear flow is interpreted as the flow from NFs (fed by the reservoir matrix) to the horizontal well [8]. The effective permeability that is obtained from the analysis of this linear flow is representative of the overall permeability of the fracture and matrix system of the dual-porosity reservoir.

Equations 3.35 and 3.36 below represent the amount of pressure-drop due to the production in these wells in oil and gas condensate URs, respectively. It should be noted that the following equations are similar to Eqs. 3.35 and 3.36 for wells with ICHFs, respectively, except frac half-length (L_f) is replaced by HW half-length (L).

$$\frac{\Delta P(t)}{q_o B_o + q_w B_w + q_g B_g} = \frac{4.064\pi\sqrt{t}}{2hn_{hf}L\sqrt{k_{eff}\lambda_t}\sqrt{(c_t\varphi)_{f+m}}} + \frac{141.2S_{hf}^{face}}{hn_{hf}k_{eff}\lambda_t} \qquad (3.35)$$

$$\frac{\Delta P(t)}{q_w B_w + \left(380\frac{\rho_o}{M_o}q_o + q_g\right)B_g} = \frac{4.064\pi\sqrt{t}}{2hn_{hf}L\sqrt{k_{eff}\lambda_t}\sqrt{(c_t\varphi)_{f+m}}} + \frac{141.2S_{hf}^{face}}{hn_{hf}k_{eff}\lambda_t} \qquad (3.36)$$

where S_{hw} is the amount of skin factor related to the production via the HW.

Hence, by plotting LHS terms of above equations versus $\sqrt{t}$, slope of the resulting straight-line would be Eq. 3.37, and if all parameters are known, effective

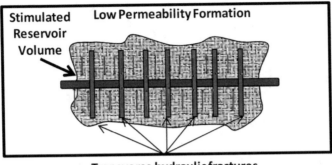

Fig. 3.9 Schematic of linear flow in a stimulated multistage hydraulically fractured HW in a dual-porosity reservoir [8]

permeability (k_{eff}) can be calculated:

$$m = \frac{1}{2hn_{\text{hf}}L\sqrt{k_{\text{eff}}\lambda_t}\sqrt{(c_t\varphi)_{\text{f+m}}}} \tag{3.37}$$

3.2.2.3 Linear Flow Analysis in Stimulated HW in a Dual-Porosity Reservoir

The general linear flow mechanisms in a horizontal well with an ICHF in a dual-porosity reservoir is demonstrated in Fig. 3.9. It should be noted that the flow relationships and the process of pressure analysis in these scenarios are similar to the previous case (linear flow in an unstimulated horizontal well in the dual-porosity reservoir). It should be emphasized here, effective permeability obtained from the straight-line analysis, unlike previous scenarios which indicated the permeability of the fracture and matrix system in the unstimulated reservoir volume, represents the permeability of the stimulated area of the reservoir.

3.2.2.4 Field Examples of Linear Flow Analysis

The following ones are two field examples of linear flow PTA, based on the study by Kurtoglu [8]. Both examples are from fractured wells that have undergone shut-in period (pressure build-up test) after a relatively long-term flow period (pressure draw-down).

The first example is a 9392-foot horizontal well with a single frac, which experiences a three-day build-up test after two months of production. Figure 3.10 shows pressure and flowrate history of this well test, as well as a pressure and pressure derivative plot. Furthermore, Fig. 3.11 is the straight-line analysis (pressure changes versus of time) of this test.

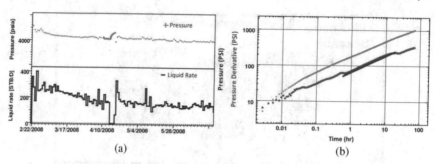

Fig. 3.10 Plot of pressure and flowrate (**a**) and plot of pressure and pressure derivative (**b**) of a build-up test in an stimulated HW in Bakken field [8]

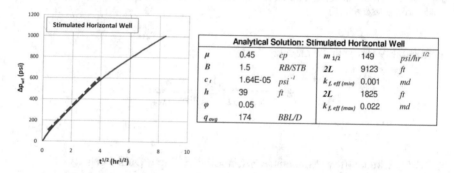

Analytical Solution: Stimulated Horizontal Well					
μ	0.45	cp	$m_{1/2}$	149	$psi/hr^{1/2}$
B	1.5	RB/STB	$2L$	9123	ft
c_t	1.64E-05	psi^{-1}	$k_{f, eff (min)}$	0.001	md
h	39	ft	$2L$	1825	ft
φ	0.05		$k_{f, eff (max)}$	0.022	md
q_{avg}	174	BBL/D			

Fig. 3.11 Linear flow analysis of stimulated horizontal build up test in Bakken field [8]

As shown in Fig. 3.10b, 0.5 is found from the slope analysis of the logarithmic pressure–time diagram. This half-slope in the logarithmic plot is itself a signature of a linear flow in the production system of the well and reservoir.

From the results shown in Fig. 3.11, it can be inferred that considering 20% of the well length as the effective length, effective permeability of the fracture is 0.022 md, and considering the entire well length, the effective permeability of the fracture would be 0.001 md.

The second example is comprised of 30 days of production followed by three days of build-up period from a fractured vertical well in Three Forks formation. Average production of this well is near 22 barrels per day. PTA plots and results of this test are depicted in Fig. 3.12 and 3.13 [8].

In this example, two interpretations have been made as follows: 80 ft frac length has an effective permeability of 0.022 md and 160 ft frac length would have an effective permeability of 0.005 md.

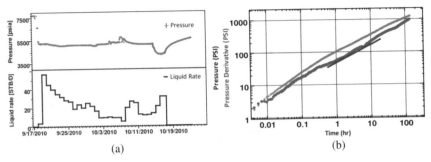

Fig. 3.12 Plot of pressure and flowrate (**a**) and plot of pressure and pressure derivative (**b**) of a build-up test in a stimulated vertical well in Three Forks field [8]

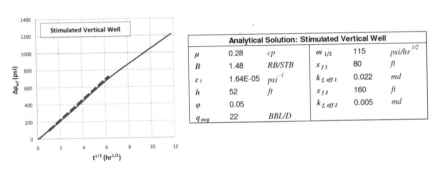

Fig. 3.13 Linear flow analysis of stimulated vertical well build-up test in Three Forks field [8]

3.2.3 Type Curve Matching Analysis

In analyzing well test data using type curves, which are mathematically derived pressure plots in logarithmic scales, pressure data that are acquired from well test are matched with type curves. Reservoir model and relevant properties can be determined using a type curve matching procedure. This method, which was initially developed in the PTA realm [5, 10] for conventional oil reservoirs, is now being adopted for a variety of flow geometries and physics in UORs/UGRs.

3.2.3.1 PTA Type Curves for MFHWs in SGRs

If flow equation from dual-porosity model of SGR while considering the effects of adsorbed gas are defined by Eq. 2.20, type curve can be acquired. In this regard, following the study by [17], boundary and initial conditions for point source analytical solution of this equation in Laplace space will be Eqs. 3.38–3.40.

Outer boundary condition is as follows:

$$\Delta \overline{m}(P_{\mathrm{f}})\big|_{r_{\mathrm{D}} \to \infty} = 0 \tag{3.38}$$

Inner boundary condition:

$$\lim_{\varepsilon \to 0^+} k_f L \left(r_D^2 \frac{\partial \Delta \overline{m}(P_f)}{\partial r_D} \right)_{r_D=\varepsilon} = -25150 \tilde{q} \delta(t) \tag{3.39}$$

Initial condition:

$$\Delta \overline{m}(P_f) \big|_{t_D=0} = 0 \tag{3.40}$$

where; $\Delta \overline{m}(P_f)$ is pseudo-pressure difference of NF in Laplace space, r_D is dimensionless radius, s is Laplace variable, λ is inter-porosity flow coefficient, ω_f is fracture storativity ratio, ω_d is desorption storativity ratio, and t_D is dimensionless time which are introduced in Table 3.2. It should be noted that Tables 3.2 and 2.3 are equivalent, though the former is in field unit and the latter in metric system. Additionally, following the solution of this equation at a single point of HF in the well and reservoir system (Fig.2.1), the inner boundary condition would indicate the pulse production in $\tilde{q}$ (cubic feet per day) at time $t = 0$ from the surface of a small sphere. The $\delta(t)$ term in Eq. 3.40 denotes the Dirac delta function.

Based on the point source solution in HFs introduced by Ozkan and Raghavan (1988) [9], and also by Zhao et al. (2013) [17], solution of the flow equation in such production system of an SGR (Eq. 2.20) in terms of pseudo-pressure is obtained as Eq. 3.41.

$$\Delta \overline{m}(P_f) = 9.4134 \times 10^{13} \frac{P_{sc}T}{T_{sc}} \frac{\overline{q}(s)}{k_f h} K_0 \left[\sqrt{f(s)} r_D \right] \tag{3.41}$$

In this equation, T is reservoir temperature, k_f is the NF permeability, $\overline{q}(s)$ is production flowrate in the Laplace space, and K is the modified Bessel function of second type.

Table 3.2 Definition of dimensionless and inter-porosity parameters for UGRs		
Dimensionless radius	$r_D = \frac{r}{L}$
Dimensionless time	$t_D = \frac{0.006328 k_f}{(\varphi_f c_{tf} + \varphi_m c_{tm})\mu L^2} t$
Dimensionless pseudo-pressure	$m_D(P) = \frac{k_f h T_{sc}}{50,300 P_{sc} q_{sc} T} \Delta m(P)$
Desorption compressibility	$c_d = \frac{P_{sc}T(1-\varphi_m-\varphi_f)\overline{Z}_m}{\varphi_m T_{sc}} \frac{V_L}{(P_L+\overline{P}_m)^2}$
Inter-porosity flow coefficient	$\lambda = \alpha \frac{k_m}{k_f} L^2, \alpha = \frac{4n(n+2)}{h_m^2}$
Fracture storativity ratio	$\omega_f = \frac{c_{tf}\varphi_f}{c_{tf}\varphi_f + c_{tm}\varphi_m}$
Matrix storativity ratio	$\omega_m = \frac{c_{tm}\varphi_m}{c_{tf}\varphi_f + c_{tm}\varphi_m} = 1 - \omega_f$
Desorption storativity ratio	$\omega_d = \frac{c_d \varphi_m}{c_{tf}\varphi_f + c_{tm}\varphi_m}$

The definition in Eq. 3.42 can also be used for dimensionless radius.

$$r_D = \sqrt{(x_D - x_{wD})^2 + (y_D - y_{wD})^2} \qquad (3.42)$$

where x and y are Cartesian distance coordinates.

Now one can use the definition of dimensionless pseudo-pressure in Table 2.1 to express the amount of dimensionless pseudo-pressure in Laplace space for each HF in such production system as follows:

$$\overline{m}_{Di}(P_f) = \frac{q_{ri}}{2L_{fDi}} \frac{1}{s} \int_{-L_{fDi}}^{+L_{fDi}} K_0\left[\sqrt{sf(s)}\sqrt{(x_D - x_{Di} - \xi)^2 + (y_D - y_{Di})^2}\right]d\xi \qquad (3.43)$$

In this equation, L_{fi} is half-length of each HF, and L_{fDi} is L_{fi} normalized to the HW half-length. Also, q_{ri} is the fraction of flowrate of each stage of HFs to the total production flowrate under standard conditions.

$$q_{ri} = \frac{q_i}{q_{sc}} \qquad (3.44)$$

In this case, by gridding the HFs as shown in Fig. 3.14, then applying the superposition principle to account for the HFs' impact on one another's pressures (Eq. 3.45), also matching the summation of the production flowrate of each frac stage with the total well flowrate (Eq. 3.46), the total well pressure response of this reservoir system can be calculated as dimensionless pseudo-pressure of the well in Laplace space ($\overline{m}_{wD}$) by solving the matrix equation written as Eq. 3.47.

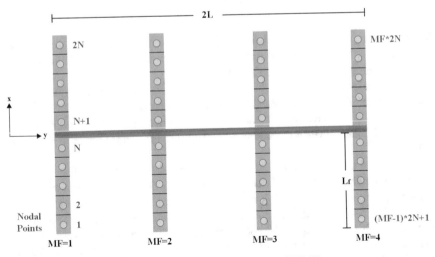

Fig. 3.14 Top view schematic of HW with four gridded stags of HF [2]

$$\overline{m}_{wD} = \overline{m}_D(x_{Dj}, y_{Dj}) = \sum_{i=1}^{NMF} \overline{m}_{Di}(x_{Dj}, y_{Dj}) \tag{3.45}$$

$$\sum_{i=1}^{NMF} q_{ri} = \frac{\sum_{i=1}^{NMF} q_i}{q_{sc}} = 1 \tag{3.46}$$

$$\begin{bmatrix} A_{1,1} & .. & A_{1,k} & .. & A_{1,n_{hf}} & -1 \\ ... & .. & ... & .. & ... & -1 \\ A_{k,1} & .. & A_{k,k} & .. & A_{k,n_{hf}} & -1 \\ ... & .. & ... & .. & ... & -1 \\ A_{n_{hf},1} & .. & A_{n_{hf},k} & .. & A_{n_{hf},n_{hf}} & -1 \\ 1 & .. & 1 & .. & 1 & 0 \end{bmatrix} \begin{bmatrix} q_{r1} \\ q_{r1} \\ . \\ . \\ q_{rn_{hf}} \\ \overline{m}_{wD} \end{bmatrix} = \begin{bmatrix} 0 \\ 0 \\ . \\ . \\ 0 \\ 1 \end{bmatrix} \tag{3.47}$$

where i is the index of HF stages and j is the index of grids considered on the frac stages. Besides, A is defined as Eq. 3.43–3.47.

$$A_{i,j} = \frac{1}{2L_{xDi}} \frac{1}{s} \int_{-L_{fDi}}^{+L_{fDi}} K_0 \left[\sqrt{sf(s)} \sqrt{(x_{Di} - x_{Dj} - \xi)^2 + (y_{Di} - y_{Dj})^2} \right] d\xi \tag{3.48}$$

To apply the effects of skin factor and wellbore storage on this pressure analysis, the relationship developed by van Everdingen and Hurst (1949) [16] is used as Eq. 3.49.

$$\overline{m}_{wD} = \frac{s\overline{m}_{wDN} + S}{s + C_D s^2 (s\overline{m}_{wDN} + S)} \tag{3.49}$$

where; $\overline{m}_{wDN}$ denotes the dimensionless pseudo-pressure in Laplace space without considering skin factor and wellbore storage, S is dimensionless skin factor, and C_D is dimensionless well storage coefficient.

Now, by taking the inverse from dimensionless pseudo-pressure in Laplace space, the amount of dimensionless pseudo-pressure in time domain can be obtained. In most PTA/RTA methods, where Laplace inversion is needed, the Stehfest (1970) [13] numerical Laplace inverse method can be used (Details can be found in Appendix A).

By plotting dimensionless pressure versus dimensionless time in a log–log scale, the corresponding type curve is found. Figure 3.15 demonstrates type curves for a dual-porosity SGR considering the effect of adsorbed gas considering the Zhao et al. model. As seen in this plot, characteristics of linear, elliptical, and radial flow regimes resulting from the well production are specified on the pressure derivative curve. MATLAB code and further explanations for plotting PTA type curve plot based on Zhao et al. model can be found in Appendix B.

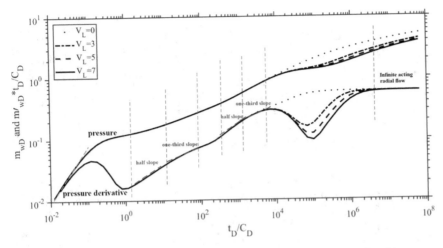

Fig. 3.15 PTA type curve plot of MFHW in the SGR with sensitivity to Langmuir volume

According to the Fig. 3.15, different flow regimes can be observed from the pressure derivative curve in graphical PTA interpretation. This type curve is plotted with sensitivity to Langmuir volume. Based on this diagram, 8 flow periods are seen in the PTA interpretation of the well and reservoir system, explained below:

- Flow period 1: only wellbore storage effect, with a unit slope characteristic for pseudo-pressure and pseudo-pressure derivatives curves, as well as the skin effect are observed. Increasing the wellbore storage coefficient causes the curve skew to the right (higher time values) and increasing the skin factor causes the curve to move upward (higher pressure values). In other words, these effects mask the reservoir and fracs impact on the PTA response of the well for a period of time.
- Flow period 2: In this flow period, the initial linear flow regime from the reservoir to each HF, based on Fig. 3.16a, with slope of 0.5 on the pseudo-pressure derivative curve is seen.
- Flow Period 3: This flow period represents the elliptical flow regime for each HF, which is characterized by a 1/3 of the slope on the pseudo-pressure derivative curve [17]. The schematic of this flow pattern is shown in Fig. 3.16b.
- Flow period 4: Theoretically, in this flow period, with the propagation of the pressure-drop wave pulse from the HF and of course in the absence of NFs in the reservoir, a radial flow regime should be observed. However, in MFHWs, due to fracs proximity to one another and their length, radial flow regime is masked. This happens because of pressure interference between these fractures [17].
- Flow period 5: In this flow period, the secondary linear flow regime is formed as shown in Fig. 3.16c and the fluid flow continues linearly to the set of HFs. This linear flow can be seen as a curve with slope of 1/2 on the pseudo-pressure derivative plot.

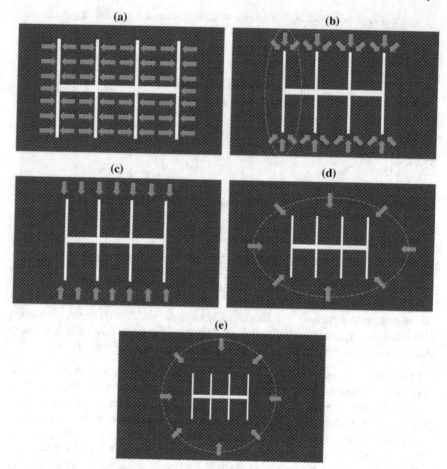

Fig. 3.16 Schematic of different flow regimes in a MFHW of an SGR [2]

- Flow period 6: After the secondary linear flow, the secondary elliptical flow, as depicted in Fig. 3.16d, is developed from the reservoir to the set of HFs. This flow regime has a slope characteristic of 1/3 on the pseudo-pressure curve.
- Flow Period 7: This period experiences the inter-porosity flow between the matrix porosity and NF. This flow is also observed in conventional dual-porosity reservoirs and is characterized by a decline in the derivative curve. This behavior represents a decrease in the rate of pressure decline during this flow period. In SGRs with adsorbed gas storage (at least in the model of Zhao et al.), flow of desorbed gas from the surface of the matrix to the NFs of the reservoir also takes place during this period, hence the desorption effect increases fluid flowrate from the matrix to the NF (reduction of pressure-drop). Ultimately, the adsorbed gas enters the flow path from the reservoir from this time on.

- Flow Period 8: Finally, with the further propagation of the pressure perturbation in the reservoir, the flow geometry reaches the flow regime of the infinite acting radial flow (IARF). During this flow interval, dimensionless pseudo-pressure derivative curve is fixed on a constant value of 0.5 (zero slope). Flow pattern of this flow period is shown in Fig. 3.16e. In this flow period, because of constant value for pseudo-pressure derivative values in the logarithmic plot and also the definition of this dimensionless pseudo-pressure derivative, reservoir permeability in the same drainage area can be calculated.

A closer look at the type curve of Fig. 3.15 and considering the sensitivity to the Langmuir volume (V_L), one can see that there is a positive correlation between the Langmuir volume and the magnitude of pseudo-pressure derivative curve's decline. This happens because in shale gas, pressure maintenance with greater desorption capacity increases during the inter-porosity flow period. The following type curves, which was initially introduced by Zhao et al., investigate the sensitivity of PTA curves to horizontal well lengths (Fig. 3.17) and the number of HFs (Fig. 3.18).

From Fig. 3.16, it can be implied that, in Zhao et al. PTA model for SGRs, HW length affects the secondary linear and elliptical flow regimes as well as the inter-porosity flow regime on the derivative curve. In addition, Fig. 3.17 exhibits that the number of HFs would affect the pressure behavior of the initial linear and elliptical flow regimes. Generally, reducing the well length as well as the number of HFs increases the pressure-drop (upward shift of the dimensionless pseudo-pressure curve) and also increases the rate of pressure decline (upward shift of the dimensionless pseudo-pressure derivative curve). Furthermore, a reduction in the length of the HW leads to an earlier onset of inter-porosity flow due to the faster propagation of the pressure-drop pulse away from the well and the HFs system.

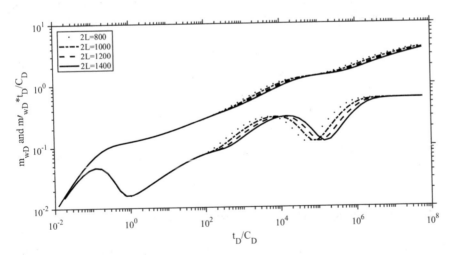

Fig. 3.17 PTA type curve plot of MFHW in the SGR with sensitivity to well length (2L)

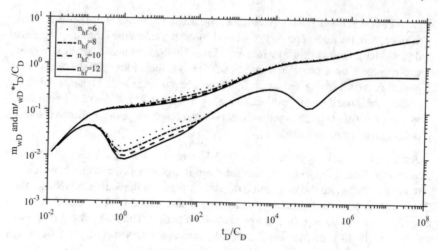

Fig. 3.18 PTA type curve plot of MFHW in the SGR with sensitivity to the number of HFs

Type curve matching, in PTA, is matching well test data from the field with a set of plotted type curves that have sensitivity to one or more parameters to determine what they are. For instance, Figs. 3.15, 3.17 and 3.18 are type curves with sensitivity to Langmuir volume, HW length, and number of HFs, respectively. In addition, calculating various properties using type curve matching, as an alternative method, one can determine key parameters such as fracture permeability or HW half-length using the definition of dimensionless parameters and also choose a matching point (where there is a match with field data) on the type curve plot. As two important examples, HW half-length and NF permeability can be calculated using Eqs. 3.50 and 3.51, respectively, after selecting a matching point from matched type curves.

$$L = \sqrt{\frac{0.006328 k_f}{(\varphi_f c_{tf} + \varphi_m c_{tm})\mu} \left(\frac{t}{t_D}\right)_M} \tag{3.50}$$

$$k_f = \frac{50,300 P_{sc} q_{sc} T}{h T_{sc}} \left(\frac{m_D(P)}{\Delta m(P)}\right)_M \tag{3.51}$$

where the terms with subscript "M" indicate the parameters obtained by matching point.

3.2.3.2 Solution for Different Reservoirs and Frac Types

HFs in terms of pressure change, are divided into three categories of uniform flux, infinite conductivity, and finite conductivity while each would affect MFHW's completion in a unique way. Uniform flux fracs have a constant flow along the fracture and ICHFs have a negligible pressure-drop along the fracture. Also, FCHFs, alike uniform flux HFs, have a noticeable amount of pressure-drop along the fracture while their flowrate is not constant. Finite conductivity fracs are comprised of major HFs in conventional wells and reservoirs, while due to very low intrinsic permeability of URs, fracs created in such reservoirs are mainly ICHFs.

Conductivity of HFs is an important parameter which not only reveals the strength of the frac, but also can determine if it is an infinite or finite conductive. Below, Eqs. 3.52 and 3.53 refer to the definition of frac conductivity and dimensionless frac conductivity, respectively.

$$F_C = k_f w_f \tag{3.52}$$

$$F_{CD} = \frac{k_f w_f}{k L_f} \tag{3.53}$$

where; k_f, $w_{f,}$ and L_f refer to permeability, width, and length of the HF, correspondingly, and k is the intrinsic permeability of the reservoir. Fractures with F_{CD} values greater than 300 are considered as ICHFs [14], and this is evident in URs, because extremely low reservoir permeabilities in micro-/nano-Darcy scale, should have F_{CD} values much larger than 300.

Below, as originally presented by Restrepo (2008) [11], general pressure responses resulting from flow through wells with transverse fracs, considering a homogenous host reservoir (non-fractured) or naturally fractured, is given. It should be noted that the solutions in Table 3.3 are independent of reservoir fluid phase (oil or gas), although they are based on a single-phase fluid assumption. In fact, solution for oil reservoirs can be expressed in Laplace space dimensionless pressure ($\overline{P}_D$) and for gas reservoirs in Laplace space dimensionless pseudo-pressure ($\overline{m}_D$).

Where in Table 3.3, for finite conductivity fracs, P_{wD} is dimensionless bottomhole pressure and $\overline{q}$ denotes the fluid flowrate at HF plane obtained from dimensionless pressure gradient in porous media, defined as:

$$\overline{q}(\xi, s) = -\frac{2}{\pi} \frac{\partial \overline{P}_D}{\partial y_D}\bigg|_{y_D = w_{fD}} \tag{3.54}$$

For instance, previously introduced model by Zhao et al. took advantage of the pressure solution in uniform flux HFs in naturally fractured SGR (Eqs. 3.41 and 3.43). However, by considering all pressure-drops in HF grids and the well, equal, infinite conductivity frac pressure behavior was achieved. Nonetheless this is not the only approach to account for ICHFs in PTA models. As shown in Table 3.3, the pressure response for the ICHF is equal to the equivalent of pressure response from the uniform flow fracture at 0.732 of the length of both frac half-lengths—($x_D = 0.732$, $y_D = 0$) or ($x = 0.732L_f$, $y = 0$). This means at this specific point over the fracture length, the flowrate of a ICHF and a uniform flux frac has become similar,

Table 3.3 Laplace space pressure response of fractured wells for various HF/reservoir types

Dimensionless pressure response in Laplace space	Frac type
Homogeneous reservoir	
Uniform flux	$\overline{P}_D(x_D, y_D, s) = \frac{1}{2s} \int\limits_{-1}^{+1} K_0\left[\sqrt{s}\sqrt{(x_D - \xi)^2 + y_D^2} \right] d\xi$
Infinite conductivity	$\overline{P}_D(s) = \frac{1}{2s} \int\limits_{-1}^{+1} K_0\left[\sqrt{s}\sqrt{(0.732 - \xi)^2} \right] d\xi$
Finite conductivity	$\overline{P}_{wD}(x_D, s) = \frac{1}{2} \int\limits_{0}^{1} \overline{q}(\xi, s)\left\{ K_0\left[\sqrt{s}\sqrt{(x_D - \xi)^2} \right] + K_0\left[\sqrt{s}\sqrt{(x_D + \xi)^2} \right] \right\} d\xi -$ $\frac{\pi}{F_{CD}} \int\limits_{0}^{x_D} \int\limits_{0}^{\xi} \overline{q}(\xi', s) d\xi' d\xi + \frac{\pi x_D}{F_{CD}s}$
Naturally fractured reservoir	
Uniform flux	$\overline{P}_D(x_D, y_D, s) = \frac{1}{2s} \int\limits_{-1}^{+1} K_0\left[\sqrt{sf(s)}\sqrt{(x_D - \xi)^2 + y_D^2} \right] d\xi$
Infinite conductivity	$\overline{P}_D(s) = \frac{1}{2s\sqrt{sf(s)}}\left[\int_0^{\sqrt{sf(s)}(1+0.732)} K_0(\xi) d\xi + \int_0^{\sqrt{sf(s)}(1-0.732)} K_0(\xi) d\xi \right]$
Finite conductivity	$\overline{P}_{wD}(x_D, s) =$ $\frac{1}{2} \int\limits_{0}^{1} \overline{q}(\xi, s)\left\{ K_0\left[\sqrt{sf(s)}\sqrt{(x_D - \xi)^2} \right] + K_0\left[\sqrt{sf(s)}\sqrt{(x_D + \xi)^2} \right] \right\} d\xi -$ $\frac{\pi}{F_{CD}} \int\limits_{0}^{x_D} \int\limits_{0}^{\xi} \overline{q}(\xi', s) d\xi' d\xi + \frac{\pi x_D}{F_{CD}s}$

consequently the pressure change will become identical. In Fig. 3.19, a schematic of similarity of flowrates of these two types of fracs is shown at 0.732 ratio of the frac half-length. In this figure, the flow profile of these two types of HFs is depicted along the frac (two half-lengths), which is constant in relation to the uniform flux frac, and a nonlinear curve to the infinite transmission ICHF. It is understood that at 0.732 as the ratio of both half-lengths of the frac, the flowrates from either of the two HF types would be the same.

As mentioned earlier, in oil reservoirs, each of the pressure responses is in the form of dimensionless pressure only containing oil. In addition to the pressure term, other parameters used in these formulations have to be in the context of oil reservoirs. By considering all parameters used in UORs PTA relationships, all would be the same as those introduced in Table 3.2 except dimensionless pressure, which is defined as:

Fig. 3.19 Flow profiles along with uniform flux and infinite conductivity fracs (Modified from [6])

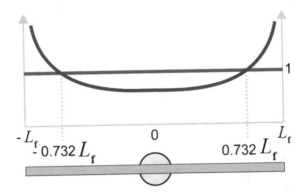

$$P_D = \frac{k_f h (P_i - P_{wf})}{141.2 q_{sc} \mu} \tag{3.55}$$

Also, for a UOR with no sorption effect $f(s)$ term is:

$$f(s) = \frac{\lambda + \omega_f (1 - \omega_f) s}{\lambda + (1 - \omega_f) s} \tag{3.56}$$

PTA type curves for wells with uniform flux and infinite conductivity transverse fracs in both homogenous and naturally fractured reservoirs (NFRs), concerning the solutions in Table 3.3, are depicted in Figs. 3.20 and 3.21. In these pressure and pressure derivative type curves, effects of wellbore storage and skin are not included. Besides, in NFRs, storativity ratio and inter-porosity flow coefficient are considered as $\omega = 0.01$ and $\lambda = 100$, respectively.

3.2.3.3 Are Post-Frac Well Tests in URs Feasible in Real-World?

Application of post-frac PTA in URs focuses on the analysis of initial linear flows and reservoir evaluation through pressure data. Hence, at later times (infinite acting radial flow) this data cannot be used in practice. In fact, propagation of the pressure-drop pulse in such low-pressure reservoirs entails a significant amount of time for build-up test (long-term well shut-in) and draw-down test (long-term production with the constant flowrate). However, dedicating a large period of time to well test does not have economic justifications and is practically impossible. Figure 3.22 depicts a common PTA of URs on a pressure–time log–log plot. Based on this figure, investigation of linear flow can take up to approximately 30 years (10,000 days) of production from the reservoir, and the analysis of subsequent times requires hundreds to thousands of years of production from a single-well in the UR which is never operationally feasible [6]. Therefore, post-frac PTA techniques which are time-consuming even for analyzing early linear flows in MFHWs, are usually replaced with RTA techniques

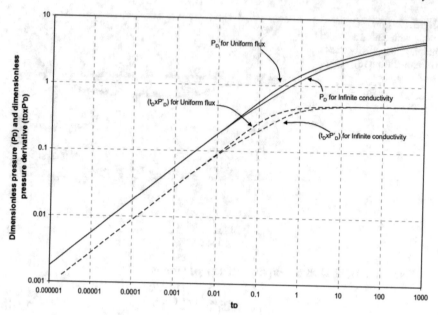

Fig. 3.20 PTA type curves for a well with different vertical frac types in a homogeneous reservoir [11]

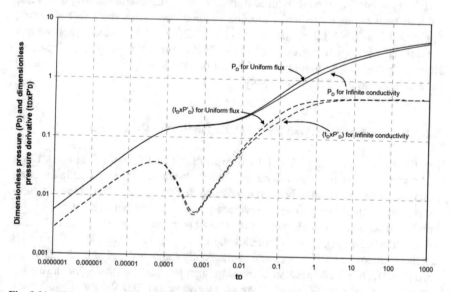

Fig. 3.21 PTA type curves for a well with different vertical frac types in an NFR [11]

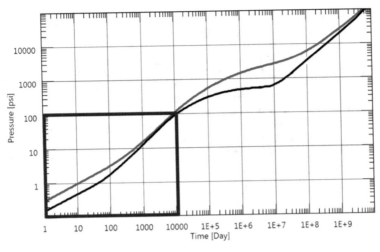

Fig. 3.22 Operational time range for PTA analysis in URs depicted on a sample pressure–time plot [6]

in the field, to achieve reservoir evaluation not only thorough a well test operation, but also during daily production from the wells.

References

1. Baba A, Azzouguen A, Mazouzi A, et al (2002) Determination of the controlling factors and origins of the bilinear flow from horizontal well transient responses. In: Canadian international petroleum conference (CIPC). Calgary, Alberta
2. Dahim S, Taghavinejad A, Razghandi M, et al (2020) Pressure and rate transient modeling of multi fractured horizontal wells in shale gas condensate reservoirs. J Pet Sci Eng 185:106566. https://doi.org/10.1016/j.petrol.2019.106566
3. Du KF, Stewart G (1992) Transient pressure response of horizontal wells in layered and naturally fractured reservoirs with dual-porosity behavior. In: SPE annual technical conference and exhibition. society of petroleum engineers, Washington, DC
4. Economides MJ, Nolte KM (1987) Reservoir stimulation Handbook. Schlumberger, Houston, TX
5. Gringarten AC, Ramey HJ, Raghavan R (1974) Unsteady-state pressure distributions created by a well with a single infinite-conductivity vertical fracture. SPE J 14:347–360. https://doi.org/10.2118/4051-pa
6. Houze O, Viturat D, Fjaere OS (2018) Dynamic data analysis. V 5.20. 01. Kappa Engineering
7. Joseph JA, Koederitz LF (1985) Unsteady-state spherical flow with storage and skin. SPE J 25:804–822. https://doi.org/10.2118/12950-PA
8. Kurtoglu B (2013) Integrated reservoir characterization and modeling in support of enhanced oil recovery for bakken. Colorado School of Mines
9. Ozkan E, Raghavan R (1988) Performance of horizontal wells. University of Tulsa
10. Ramey HJ (1970) Short-time well test data interpretation in the presence of skin effect and wellbore storage. J Pet Technol 22:97–104. https://doi.org/10.2118/2336-pa
11. Restrepo DP (2008) Pressure behavior of a system containing multiple vertical fractures. University of Oklahoma

12. Rezaee R (2015) Fundamentals of gas shale reservoirs. Wiley Online Library
13. Stehfest H (1970) Numerical inversion of Laplace transform. Commun ACM 13:47–49. https://doi.org/10.1145/361953.361969
14. Stewart G (2011) Well Test Design & Analysis. PennWell Corporation
15. Torcuk MA, Kurtoglu B, Alharthy N, Kazemi H (2013) Analytical solutions for multiple matrix in fractured reservoirs: application to conventional and unconventional reservoirs. SPE J 18:969–981. https://doi.org/10.2118/149501-MS
16. van Everdingen AF, Hurst W (1949) The application of the Laplace transformation to flow problems in reservoirs. J Pet Technol 1:305–324. https://doi.org/10.2118/949305-G
17. Zhao Y, Zhang L, Zhao J et al (2013) "Triple porosity" modeling of transient well test and rate decline analysis for multi-fractured horizontal well in shale gas reservoirs. J Pet Sci Eng 110:253–262

Chapter 4
Rate Transient Analysis

4.1 Decline Rate Analysis

Naturally, by producing from the reservoirs due to reduction in reservoir average pressure and the resulting change in production mechanisms, the rate of production will experience a decreasing trend over time. This declining trend in production rate is not that significant in the development phase of the fields when new wells are drilled and come into production. However, as production continues, this declining trend in the production rate from the wells become inevitable.

Initially, Arps in 1945 [2] experimentally modeled the declining trend in production flowrate for wells in conventional reservoirs and presented his decline curves to predict future production from the reservoir, enabling us to determine production mechanisms, and estimate ultimate recovery (EUR) of the reservoir. Likewise, similar methods for decline rate analysis of the wells in unconventional reservoirs (URs) is presented, and discussed in this chapter.

4.1.1 Arps Decline Curve Analysis

Based on experimental observations by Arps [2], if operational conditions such as bottomhole pressure (BHP) stays unchanged over time, Eq. 4.1 can define decline curve analysis (DCA) of the well during the reservoir drainage period. Additionally, Arps' equation defines production rate during boundary-dominated flow (BDF) and not transient flow. It should be noted that BDF is a general concept which covers pseudo-steady state (PSS) flow as one its components. This means PSS flow regime can only be defined if production flowrate is constant and subsequently under limited reservoir boundary conditions. However, when production control mode is changed to constant BHP, the concept of PSS flow regime is no longer valid, and general BDF concept will become the dominant flow regime.

A. Taghavinejad et al., *Unconventional Reservoirs: Rate and Pressure Transient Analysis Techniques*, SpringerBriefs in Petroleum Geoscience & Engineering, https://doi.org/10.1007/978-3-030-82837-0_4

$$q(t) = q_i(1 + bD_it)^{\frac{-1}{b}}$$

(4.1)

where $q(t)$ is the rate at time t, b is the rate exponent (dimensionless), and D_i is the nominal rate decline at time zero (1/day).

Using Eq. 4.1 and assigning q_i, b and D_i to the studied well, assuming that the operating conditions of the well and reservoir are unchanged, time-dependent declining trend of rate can become predictable. Considering experimental observations by Arps and also three separate values for b, three separate types of Arps decline rate curves can be defined as follows: $b = 0$, exponential decline rate; $b = 1$ harmonic decline rate; and $0 < b < 1$, is hyperbolic decline rate. Figure 4.1 is the decline rate vs. time plot for a conventional oil reservoir depicting the differences between these three types of decline rate curves.

Considering the curves in Fig. 4.1, it is evident that exponential curve has the lowest decline rate (q) and the harmonic one the highest. As a result, it appears that larger b values would indicate a stronger production from the reservoir with higher rates while lower rate should have smaller b values. Furthermore, in hyperbolic decline rate, very weak and very strong decline rate happens when b varies between 0 and 1 ($0 < b < 1$).

Following a mathematical manipulation for the exponential, hyperbolic and harmonic rate declines, time-dependent rate, $q(t)$, time-dependent cumulative production, $N_p(t)$, and ultimate cumulative production, $N_p(\infty)$, for each would become as those summarized in Table 4.1. It should be noted that cumulative produc-

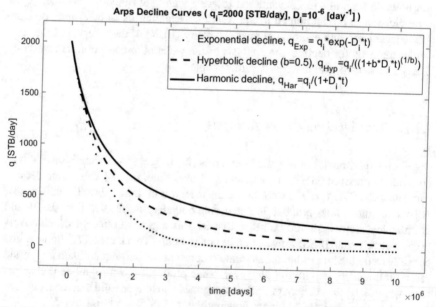

Fig. 4.1 Three separate forms of Arps decline rate curves

Table 4.1 Rate and cumulative production relations of Arps decline model

Rate production	Cumulative production	Final cumulative production
Exponential rate decline (b = 0)		
$q(t) = q_i \exp(-D_i t)$	$N_p(t) = \frac{q_i}{D_i}\left(1 - e^{-D_i t}\right)$	$N_p(\infty) = \frac{q_i}{D_i}$
Hyperbolic rate decline (0 < b < 1)		
$q(t) = q_i(1 + bD_i t)^{\frac{-1}{b}}$	$N_p(t) =$ $\frac{q_i}{(b-1)D_i}\left(1 - e^{-D_i t}\right)\left[1 + bD_i t^{\left(1-\frac{1}{b}\right)} - 1\right]$	$N_p(\infty) =$ $\frac{q_i}{D_i}\left(\frac{1}{1-b}\right)$
Harmonic rate decline (b = 1)		
$q(t) = q_i(1 + D_i t)^{-1}$	$N_p(t) = \frac{q_i}{D_i}\ln(1 + D_i t)$	$N_p(\infty) = \infty$
Beyond hyperbolic rate decline (b ≥ 1)		
$q(t) = q_i(1 + bD_i t)^{\frac{-1}{b}}$	$N_p(t) =$ $\frac{q_i}{(b-1)D_i}\left(1 - e^{-D_i t}\right)\left[1 + bD_i t^{\left(1-\frac{1}{b}\right)} - 1\right]$	$N_p(\infty) = \infty$

tion values are obtained by integrating the rate formulae respect to time whereas ultimate cumulative production values are obtained when time approaches infinity in time-dependent cumulative production formulae.

In most case scenarios and during the life of a well, since operating conditions and production mechanisms vary, b and D will inevitably change as well. This means to match production history that may not fit an ideal Arps equation (constant production conditions and flow in PSS period), several ideal Arps equations (fixed b and D values) should be used for different periods of production [16]. For example in shale reservoirs; b, in the first few days of production from the well, is about 4, it reaches 2 for a few weeks up to a few months, and it will eventually approach to zero, which contradicts the trend that is seen in conventional reservoirs when initial production is represented with $b = 0$ equivalent to the exponential decline rate, and after a few days, between zero and one (hyperbolic decline rate) [23].

Also, the initial decline rate (D_i), also known as the nominal decline rate, won't have a steady trend over time, except during exponential decline rate, thus, time-dependent rate of decline can be expressed as Eq. 4.2 [23].

$$D(t) = -\frac{d \ln(q)}{dt} = \frac{D_i}{1 + bD_i t} \tag{4.2}$$

Another valuable parameter is the effective decline rate (d), which is defined by Eq. 4.3.

$$d = \frac{q(t_1) - q(t_2)}{q(t_1)} \tag{4.3}$$

Equations 4.2 and 4.3 can be related to one another by:

$$d = 1 - e^{-D}$$

(4.4)

Finally, considering Eq. 4.5, cumulative production, N_p (t), initial rate, q_i, and also rate at any moment, $q(t)$, can relate to each other as follows [23]:

$$N_p(t) = \frac{q_i}{(b-1)D_i}\left[\left(\frac{q_i}{q(t)}\right)^{(b-1)} - 1\right]$$

(4.5)

4.1.2 Unconventional Reservoirs Decline Curve Analysis

Since Arps decline rate analysis was initially developed for conventional reservoirs and was improperly used in URs which provides erroneous results, it would be necessary to modify Arps DCA specific to URs. As it was explained earlier, the rate of decline in shale reservoirs has three flow periods: $b = 4$, $b = 2$, and then b approaching to zero. Hence, in shale reservoirs, bD_it will be sufficiently greater than one, subsequently, Eq. 4.6 can be replaced by Eq. 4.1.

$$q(t) = q_i(bD_i)^{\frac{-1}{b}}t^{\frac{-1}{b}}$$

(4.6)

This Equation can also be rearranged as:

$$\frac{1}{q(t)} = \frac{(bD_i)^{\frac{1}{b}}}{q_i}t^{\frac{1}{b}}$$

(4.7)

Now, by considering all three values that b can take for different flow periods in a shale reservoir that is producing with a multi-stage fractured horizontal well (MFHW), the effects of various flow regimes (bilinear, linear, boundary-dominated flow regimes) can be developed [23].

4.1.2.1 Bilinear Flow Regime of URs' Decline Rate Curves

By considering $b = 4$, bilinear flow equation in a vertical hydraulic fracture (HF) is obtained as:

$$\frac{1}{q(t)} = \frac{(4D_i)^{\frac{1}{4}}}{q_i}t^{\frac{1}{4}}$$

(4.8)

Thus, in a bilinear flow regime the slope of the logarithmic plot of rate versus time with the value of -0.25 can properly represent such flow regime.

4.1.2.2 Linear Flow Regime of URs' Decline Rate Curves

For $b = 2$, the linear flow equation in a vertical fracture results in Eq. 4.9.

$$\frac{1}{q(t)} = \frac{(2D_i)^{\frac{1}{2}}}{q_i} t^{\frac{1}{2}} \tag{4.9}$$

The curve of decline rate under linear flow regime is characterized by a slope of -0.5 value in the logarithmic plot of rate versus time.

4.1.2.3 Boundary-Dominated Flow Regime of URs' Decline Rate Curves

To study the long-term rate analysis of URs, $b \leq 1$ is considered in Eq. 4.7. Thus, to achieve the production rate expression in these reservoirs under BDF conditions, Eq. 4.10 is used.

$$\frac{1}{q(t)} = \left[\frac{(bD_i)^{\frac{1}{b}}}{q_i} t^{(\frac{1}{b}-\frac{1}{2})} \right] \sqrt{t} \tag{4.10}$$

Hence, by using the above flow relationship and plotting the inverse flowrate data ($1/q$) versus square toot of time ($\sqrt{t}$), and considering b decreasing with time, an increase in the slope should be expected. This specific characteristic of BDF with two other mentioned flow regimes (bilinear and linear) is observed in real field flowrate data plotted versus time presented in Fig. 4.2.

It should be noted that all of these separate DCAs of bilinear, linear, and boundary-dominated is still based on empirical observations while in the next section, the flow rate will be discussed analytically which is relating DCA to the rate transient analysis (RTA).

Another important point in the analysis of decline rate curves is the calculation of the EUR of the reservoir. To do so, first, since we learnt that b tends to approach to zero eventually, cumulative production of the BDF period is calculated from cumulative production formulae of exponential decline rate (shown in Table 4.1) with $b = 0$. Then by adding cumulative production rate from previous flow periods to the BDF's, EUR can be obtained.

Hence, the cumulative production of BDF period can be calculated as Eq. 4.11.

$$N_p(t - t_2) = \frac{q_2}{D_2} \left[\exp(-D_2 t_2) - \exp(-D_2 t) \right] \tag{4.11}$$

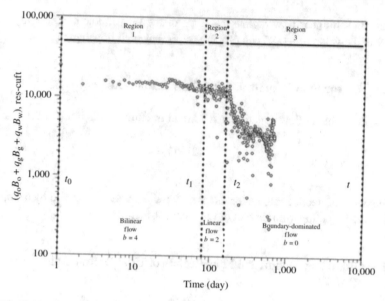

Fig. 4.2 Field data for DCA of a production well in Eagle Ford field [23]

where subscript "2" represents the onset of the flow period under BDF conditions, and t marks the time when EUR calculation is desired.

To calculate EUR, it is sufficient to calculate the cumulative production from the initial time (t_0) to the onset of BDF period (t_2) and then add that to the cumulative production in Eq. 4.11 as:

$$EUR = N_p(t - t_0) = N_p(t_2 - t_0) + N_p(t - t_2) \qquad (4.12)$$

Cumulative production from t_0 to t_2, also is known as $N_p(t_2 - t_0)$, can be calculated for unconventional oil reservoirs (UORs) and unconventional gas reservoirs (UGRs) as Eqs. 4.13 and 4.14, respectively [23].

$$N_p(t_2 - t_0) = SRV.n_{hf}\left[\left(\varphi\frac{S_o}{B_o}\right)_0 - \left(\varphi\frac{S_o}{B_o}\right)_2\right] \qquad (4.13)$$

$$G_p(t_2 - t_0) = SRV.n_{hf}\left[\left(\varphi\frac{S_g}{B_g}\right)_0 - \left(\varphi\frac{S_g}{B_g}\right)_2\right] \qquad (4.14)$$

In these equations; SRV is the reservoir volume stimulated by the HF stages (stimulated reservoir volume), n_{hf} is the number of frac stages, and φ is the porosity of the formation. Moreover, S denotes the saturation of oil and gas, B is the formation volume factor of oil and gas and also subscripts "0" and "2" refer to the relevant properties at each time values of t_0 and t_2, respectively.

4.2 Rate Transient Analysis

Rate transient analysis (RTA) methods have been developed in recent decades for analyzing production data that, unlike experimental methods of analyzing rate of decline, are based on analytical (or numerical) solution of the fluid flow equation in the reservoir. In some of these methods, complex operational conditions of the well and the reservoir such as interval well open and shut-in, and changing BHP conditions are considered as well. They are also referred to as advanced methods of analyzing the production data while DCA is known as traditional production data analysis techniques. In this book, primary RTA methods, which were initially developed for conventional reservoirs are briefly reviewed and then RTA methods specific to URs are discussed.

4.2.1 Primary Rate Transient Analysis Techniques

RTA is a method of analyzing rate-time data, which helps to determine particular well-reservoir properties as permeability, skin factor, the amount of fluid in place in the wellbore drainage area, as well as other information from the reservoir such as the near wellbore, reservoir, and reservoir boundary flow model. Alike PTA models, RTA techniques are based on various methods of data handling—here production rate—including type curve matching and straight-line analysis. In the following sections both of these approaches are briefly explained for conventional reservoirs RTA.

4.2.1.1 Primary RTA Type Curves

A good understanding of the governing flow equation in a conventional oil reservoir with a vertical well in a dimensionless state, would be the primary step to develop RTA formulae, following Eq. 4.15.

$$\frac{\partial^2 P_D}{\partial r_D^2} + \frac{1}{r_D}\frac{\partial P_D}{\partial r_D} = \frac{\partial P_D}{\partial t_D} \tag{4.15}$$

where P_D is dimensionless pressure, r_D is dimensionless radius and t_D is dimensionless time.

By applying boundary conditions of constant pressure or constant rate at the wellbore to the above equation, different answers to the dimensionless pressure will be obtained. Considering well test relationships, Eq. 4.16 represents the late-time approximate solution of the above flow equation at the wellbore (P_{wD}) with the assumption of constant flowrate at wellbore and also considering reservoir boundary to be infinite acting [24].

$$P_{wD}(t_D) = \frac{1}{2}[\ln(t_D) + 0.80908]$$ (4.16)

Equation 4.17 also shows the late-time approximate solution of Eq. 4.15 under wellbore constant flowrate and BDF conditions [24].

$$P_{wD}(t_D) = \frac{2t_D}{r_{eD}^2} + \ln(r_{eD}) - 0.75$$ (4.17)

where r_{eD} is the dimensionless outer radius of the reservoir (r_e/r_w).

By investigating the above dimensionless flow equation in either constant rate or constant bottomhole flowing pressure (BHFP) more fundamentally in the entire time (early to late times), a general response could be obtained. As mentioned earlier in Chap. 3, solving flow equation in Laplace space, in time domain using Laplace inverse methods, Eqs. 4.18 and 4.19, can be obtained under constant rate at the wellbore under infinite acting, and closed reservoir boundary conditions assumptions, respectively [24].

$$\overline{P}_D(s) = \frac{K_0(r_D\sqrt{s})}{s\sqrt{s}K_1(\sqrt{s})}$$ (4.18)

$$\overline{P}_D(s) = \frac{K_1(r_{eD}\sqrt{s})I_0(r_D\sqrt{s}) + I_1(r_{eD}\sqrt{s})K_0(r_D\sqrt{s})}{s\sqrt{s}[K_1(\sqrt{s})I_1(r_{eD}\sqrt{s}) - K_1(r_{eD}\sqrt{s})I_1(\sqrt{s})]}$$ (4.19)

In these equations; $\overline{P}_D(s)$ is dimensional pressure in Laplace space and s is the Laplace variable. Also, I_0 and I_1 are modified Bessel functions of the first kind and K_0 and K_1 are the modified Bessel functions of the second kind.

Now one can use the principle proposed by [27], to convert dimensionless pressure in Laplace space with the assumption of constant rate at the wellbore to dimensionless flowrate in Laplace space with the assumption of constant BHFP, following Eq. 4.20.

$$\overline{q}_D = \frac{1}{\overline{P}_D s^2}$$ (4.20)

where q_D is defined as the reverse of P_D ($q_D = \frac{1}{P_D}$).

Thus, the rate solution of Eq. 4.15 for a reservoir with infinite acting and a closed reservoir boundary conditions, assuming constant BHFP, is expressed as Eqs. 4.21 and 4.22, respectively.

$$\overline{q}_D(s) = \frac{K_1(\sqrt{s})}{\sqrt{s}K_0(r_D\sqrt{s})}$$ (4.21)

$$\overline{q}_D(s) = \frac{K_1(\sqrt{s})I_1(r_{eD}\sqrt{s}) - K_1(r_{eD}\sqrt{s})I_1(\sqrt{s})}{\sqrt{s}[K_1(r_{eD}\sqrt{s})I_0(r_D\sqrt{s}) + I_1(r_{eD}\sqrt{s})K_0(r_D\sqrt{s})]}$$ (4.22)

Table 4.2 Essential parameters for plotting RTA type curves of conventional oil reservoirs

Rate/pressure	Time	Rate/pressure integral	Rate/pressure
Fetkovich method [10]			
$q_{Dd} = q_D\left[\ln(r_{eD}) - \frac{1}{2}\right]$	$t_{Dd} = \dfrac{t_D}{\frac{1}{2}\left[(r_{eD})^2 - 1\right]\left[\ln(r_{eD}) - \frac{1}{2}\right]}$	$q_{Ddi} = \dfrac{N_{pDd}}{t_{Dd}}$	$q_{Ddid} = -\dfrac{dq_{Ddi}}{d\ln t_{Dd}}$
Blasingame method [22]			
$q_{Dd} = q_D\left[\ln(r_{eD}) - \frac{1}{2}\right]$	$t_{cDd} = \dfrac{N_{pDd}}{q_{Dd}}$	$q_{Ddi} = \dfrac{N_{pDd}}{t_{Dd}}$	$q_{Ddid} = -\dfrac{dq_{Ddi}}{d\ln t_{Dd}}$
Agarwal-Gardner method [1]			
$q_D = \dfrac{1}{P_D}$	$t_{DA} = \dfrac{t_D}{\pi\left(r_{eD}^2 - 1\right)}$	$\dfrac{1}{DERI} = \dfrac{1}{\frac{\partial P_{Di}}{\partial \ln t_{DA}}}$	$\dfrac{1}{DER} = \dfrac{1}{\frac{\partial P_D}{\partial \ln t_{DA}}}$
NPI method [3, 26]			
P_D	$t_{DA} = \dfrac{t_D}{\pi\left(r_{eD}^2 - 1\right)}$	$P_{Di} = \dfrac{\int_0^{t_{DA}} P_D dt_{DA}}{t_{DA}}$	$P_{Did} = -\dfrac{dP_{Di}}{d\ln t_{DA}}$

Now, using Laplace inverse methods, it would be possible to reverse the flowrate expressions to time domain. Under such conditions, the BHFP is assumed to be constant which is the basic assumption of RTA model proposed by Fetkovich (1980) [10]. It should be noted that in addition to the Fetkovich method, other methods that are described in the following sections take advantage of Eqs. 4.21 and 4.22 as their basic assumptions of fluid flow.

As premier RTA techniques applicable to conventional reservoirs including Fetkovich [10], Blasingame [22], Agarwal-Gardner [1], and NPI (normalized pressure integral) [3] methods, Table 4.2, following [26], summarizes all of the rate and time parameters that are required for plotting corresponding type curves (TCs) in oil reservoirs.

For gas reservoirs, four mentioned primary RTA methods can be used for plotting their TCs using compiled parameters in Table 4.4.

Where in Tables 4.2 and 4.3, q_{Dd} is dimensionless decline flowrate, q_{Ddi} is dimensionless decline flowrate integral, q_{Ddid} is dimensionless decline flowrate integral derivative, $1/DER$ is derivative inverse, $1/DERI$ is integral derivative inverse, P_{Di} is dimensionless pressure integral, P_{Did} is dimensionless pressure integral derivative, $m_{Di}(P)$ is dimensionless pseudo-pressure integral, $m_{Did}(P)$ is dimensionless pseudo-pressure integral derivative, and t_{Dd} is dimensionless decline time.

Also, t_c denotes the material balance time (MBT) which is responsible for making constant rate and constant BHFP solution of the governing flow equation alike their dimensionless form. In this regard, by dividing cumulative production to flowrate of the well, equivalent constant rate of time will be yielded for a set of constant BHFP time data. Hence, constant rate assumption relationships can be used for RTA of wells with constant BHFP constraint. To calculate MBT, one can use Eqs. 4.23. Eqs. 4.23 or 4.24 of its dimensional or dimensionless form.

Table 4.3 Essential parameters for plotting RTA type curves of conventional gas reservoirs

Rate/pressure function	Time function	Rate/pressure integral function	Rate/pressure derivative function
Fetkovich method [10]			
$q_{Dd} = q_D\left[\ln(r_{eD}) - \frac{1}{2}\right]$	$t_{aDd} = \dfrac{t_{aD}}{\frac{1}{2}\left[(r_{eD})^2 - 1\right]\left[\ln(r_{eD}) - \frac{1}{2}\right]}$	$q_{Ddi} = \dfrac{\int\limits_0^{t_{aDd}} q_{Dd}(\tau)d\tau}{t_{aDd}}$	$q_{Ddid} = -\dfrac{dq_{Ddi}}{d\ln t_{aDd}}$
Blasingame method [22]			
$q_{Dd} = q_D\left[\ln(r_{eD}) - \frac{1}{2}\right]$	$t_{caDd} = \dfrac{t_{caD}}{\frac{1}{2}\left[(r_{eD})^2 - 1\right]\left[\ln(r_{eD}) - \frac{1}{2}\right]}$	$q_{Ddi} = \dfrac{\int\limits_0^{t_{aDd}} q_{Dd}(\tau)d\tau}{t_{aDd}}$	$q_{Ddid} = -\dfrac{dq_{Ddi}}{d\ln t_{aDd}}$
Agarwal-Gardner method [1]			
$q_D = \dfrac{1}{m_D(P)}$	$t_{aDA} = \dfrac{t_{aD}}{\pi(r_{eD}^2 - 1)}$	$\dfrac{1}{DERI} = \dfrac{1}{\frac{\partial m_{Di}(P)}{\partial \ln t_{aDA}}}$	$\dfrac{1}{DER} = \dfrac{1}{\frac{\partial m_{Di}(P)}{\partial \ln t_{aDA}}}$
NPI method [3, 26]			
$m_D(P)$	$t_{aDA} = \dfrac{t_{aD}}{\pi(r_{eD}^2 - 1)}$	$m_{Di}(P) = \dfrac{\int\limits_0^{t_{aDA}} m_D(P)dt_{aDA}}{t_{aDA}}$	$m_{Did}(P) = \dfrac{dm_{Di}(P)}{d\ln t_{aDA}}$

$$t_c = \frac{N_p}{q} = \frac{\int\limits_0^t q(\tau)d\tau}{q} \tag{4.23}$$

$$t_{cD} = \frac{N_{pD}}{q_D} = \frac{\int\limits_0^{t_D} q_D(\tau)d\tau}{q_D} \tag{4.24}$$

Dimensionless parameters introduced in Tables 4.2 and 4.3, and also pseudo properties related to gas reservoirs used in Table 4.3 formulations are defined as follows in Table 4.4.

TC plots of all Fetkovich, Blasingame, Agarwal-Gardner, and NPI methods are depicted in Fig. 4.3. By matching field data with each of these TC plots, determining a matching point, and using dimensionless parameters introduced in Tables 4.2, 4.3, and 4.4, regarding the TC matching criteria described in Chap. 3, one can determine well/reservoir properties.

Table 4.4 Dimensionless parameters and pseudo properties used in RTA models of conventional oil/gas reservoirs

Parameter	Formulation
Oil reservoir	
Dimensionless flowrate	$q_D = \frac{141.2q\mu B}{kh(P_i - P_{wf})}$
Dimensionless pressure	$P_D = \frac{kh(P_i - P_{wf})}{141.2q\mu B}$
Dimensionless time	$t_D = \frac{0.00634kt}{\mu c_t \varphi r_w^2}$
Gas reservoir	
Dimensionless flowrate	$q_D = \frac{50,300 P_{sc} q_{sc} T}{khT_{sc}[m(P_i) - m(P_{wf})]}$
Dimensionless pseudo-pressure	$m_D(P) = \frac{khT_{sc}[m(P_i) - m(P_{wf})]}{50,300 P_{sc} q_{sc} T}$
Dimensionless pseudo-time	$t_{aD} = \frac{0.00634kt_a}{\mu c_t \varphi r_w^2}$
Pseudo-pressure	$m(P) = \frac{\mu_i Z_i}{P_i} \int_0^P \frac{P}{\mu(P)Z(P)} dP$
Pseudo-time	$t_a = \frac{\mu_i c_{ti}}{q} \int_0^t \frac{q}{\mu(P)c_t(P)} dt$

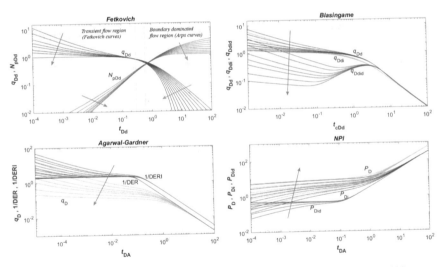

Fig. 4.3 Type curves of primary RTA methods; $r_{eD} = 10, 20, 50, 100, 200, 1000, 10,000$, $q_{Dd} = \exp(-t_{Dd})$ for Fetkovich, Blasingame, Agarwal-Gardner, and NPI TCs, and $b = 0, 0.1, 0.2, 0.3, 0.4, 0.5, 0.6, 0.7, 0.8, 0.9, 1$ for Arps curves in Fetkovich TC

4.2.1.2 Dynamic/Flowing Material Balance

Analysis of production data through dynamic/flowing material balance method is based on pressure and flow data analysis in the BDF (or PSS flow) period and determining reservoir properties such as permeability and volume of fluid in-place.

In fact, during a PSS flow period, the entire reservoir pressure decreases at a constant rate over time. At any time, average reservoir pressure (P_R) can be decided if pressure values at outer boundary of the reservoir (P_e) and also at the wellbore (P_{wf}) are known.

By considering the overall pressure-drop in the reservoir, the difference between initial pressure (P_i) and BHFP, can be separated into two distinct pressure-drop as:

$$P_i - P_{wf} = (P_R - P_{wf}) + (P_i - P_R) \tag{4.25}$$

Therefore, overall pressure-drop can be referred to the summation of average reservoir pressure difference with BHFP (P_{wf}), and initial reservoir pressure difference with average reservoir pressure.

As stated by Sun (2015) [26], each of the terms on the right hand-side of Eq. 4.25 for a reservoir in the PSS flow period are:

$$P_i - P_R = \frac{qB}{7758.36 Ahc_t\varphi} t_c \tag{4.26}$$

$$P_R - P_{wf} = \frac{70.6q\mu B}{kh} \ln\left(\frac{4A}{C_A e^\gamma r_w^2}\right) \tag{4.27}$$

where A in Eq. 4.26 is in acre-feet, and C_A in Eq. 4.27 is Dietz shape factor which is related to the shape of the reservoir drainage area, and for a circular reservoir $C_A = 4\pi e^{\frac{3}{2}-\gamma} \cong 31/62$. Additionally, γ is Euler's constant which is approximately equal to 0.577.

By embedding Eqs. 4.26 and 4.27 into Eq. 4.25, and a simple rearrangement one can have:

$$\frac{P_i - P_{wf}}{q} = b_{PSS} + mt_c \tag{4.28}$$

where;

$$m = \frac{B}{7758.36 Ahc_t\varphi} = \frac{1}{Nc_t} \tag{4.29}$$

$$b_{PSS} = \frac{70.6\mu B}{kh} \ln\left(\frac{4A}{C_A e^\gamma r_w^2}\right) \tag{4.30}$$

Similar formulations can be resulted for gas reservoirs, following [26], as:

$$\frac{\Delta m(P)}{q} = m_a t_{ca} + b_{a,PSS} \tag{4.31}$$

where;

Table 4.5 Summary of flowing material balance methods analysis outputs

	Y-axis	X-axis	Intercept	Slope
Oil reservoir				
Mattar method	$\frac{P_i - P_{wf}}{q}$	t_c	b_{PSS}	$m = \frac{1}{Nc_t}$
Agarwal-Gardner method	$\frac{q}{\Delta P}$	$\frac{N_p}{\Delta P c_t}$	$\frac{1}{b_{PSS}}$	$\frac{1}{b_{PSS} N}$
Gas reservoir				
Mattar method	$\frac{\Delta m(P)}{q}$	t_{ca}	$b_{a,PSS}$	$m_a = \frac{1}{Gc_{ti}}$
Agarwal-Gardner method	$\frac{q}{\Delta m(P)}$	$\frac{q t_{ca}}{\Delta m(P) c_t}$	$\frac{1}{b_{a,PSS}}$	$\frac{1}{G b_{a,PSS}}$

$$m_a = \frac{1}{Gc_{ti}} \tag{4.32}$$

$$b_{a,PSS} = \frac{50300(\mu B)_i}{kh}\left[\frac{1}{2}\ln\left(\frac{4A}{C_A e^{\gamma} r_w^2}\right)\right] \tag{4.33}$$

These equations are basic relationships of Mattar method of flowing material balance (FMB) [17]. In addition to this method, Agarwal-Gardner [1] is also available for FMB which can be produced by rearranging Eqs. 4.28 and 4.31 for oil and gas reservoirs, respectively.

Technically, FBM is based on straight-line analysis via cross plotting. Table 4.5 summarizes the data that is required for straight-line analysis for both oil and gas reservoirs using both Mattar and Agarwal-Gardner methods.

As it can be inferred from Table 4.5, reservoir fluid in place, N for oil reservoir and G for gas reservoir, can be estimated using the slope of the straight-line analyses. Also, reservoir permeability can be evaluated using corresponding b parameters from the intercept of the line.

4.2.2 Unconventional Reservoirs RTA Techniques

To perform RTA in URs, their specific flow assumptions must be taken into account, including the presence of MFHWs (and consequently their resulting flow regimes), and the gas sorption effect in small pores of the rock during the production. In this section, all of these contributing factors in URs' RTA will be discussed.

4.2.2.1 Bilinear Flow Analysis

It is important to note that RTA formulations can be regarded equivalent to pressure transient analysis (PTA) equations. Furthermore, to develop RTA methods such as Agarwal-Gardner and NPI where pressure relations are used for production analysis.

Table 4.6 Intercept and slope of bilinear flow straight-line RTA plots

	c	m
I	$\dfrac{141.2 S_{hf}^{well}}{h n_{hf} k_{eff} \lambda_t}$	$\dfrac{44.102}{h n_{hf} \lambda_t \sqrt{w_f k_{hf}}} \left(\dfrac{\lambda_t}{c_t \varphi k_{eff}}\right)^{\frac{1}{4}}$
II	$\dfrac{141.2 S_{hf}^{well}}{h n_{hf} k_{eff} \lambda_t}$	$\dfrac{45.103}{h n_{hf} L_f \lambda_t \sqrt{k_{eff}}} \left[\left(\dfrac{1}{1-\omega_f}\right)^{\frac{1}{2}} \dfrac{1}{\left(\frac{\alpha}{4} k_m \left(c_t \varphi \lambda_t^{-1}\right)\right)_{f+m}^{\frac{1}{4}}} \right]$
III	-	$\dfrac{45.103}{h n_{hf} L \lambda_t \sqrt{k_{eff}}} \left[\left(\dfrac{1}{1-\omega_f}\right)^{\frac{1}{2}} \dfrac{1}{\left(\frac{\alpha}{4} k_m \left(c_t \varphi \lambda_t^{-1}\right)\right)_{f+m}^{\frac{1}{4}}} \right]$

As a result, the total pressure-drop of the bilinear flow (in form of normalized to the flowrate) for different well geometries as defined in Chap. 3 can be used to develop UR's RTA relationships.

As previously discussed in Chap. 3, bilinear flow pattern has been observed in cases such as: finite-conductivity HFs (FCHFs) (I), horizontal well (HW) with infinite-conductivity HFs (ICHF) in a dual-porosity reservoir (II), and unstimulated HW in a dual-porosity reservoir (III). Available RTA plots for bilinear flow straight-line analyses, as they were initially introduced by Kurtoglu (2013) [15], are compiled in Table 4.6. There, the intercept and the slope values of the following straight-line with corresponding equations are presented.

$$\frac{\Delta P(t)}{q_o B_o + q_w B_w + q_g B_g} = c + m t^{\frac{1}{4}} \tag{4.34}$$

Using straight-line analyses, parameters like HF/formation permeability and frac/HW half-length from slope and skin factor from the intercept can be calculated in the case of bilinear flow.

4.2.2.2 Linear Flow Analysis

Alike bilinear flow analyses, straight-line RTA of linear flow is also available using their corresponding PTA formulations by taking advantage of normalized pressure with total production flowrate. Based on the study by Rezaee (2015) [23], Table 4.7 depicts the intercept and the slope of the straight-line analysis of normalized pressure equation for linear flow in ICHFs (I), and unstimulated/stimulated HW in dual-porosity reservoirs (II and III, respectively).

$$\frac{\Delta P(t)}{q_o B_o + q_w B_w + q_g B_g} = c + m \sqrt{t} \tag{4.35}$$

Table 4.7 Intercept and slope of linear flow straight-line RTA plots

	c	m
I	$\dfrac{141.2 S_{hf}^{face}}{hn_{hf}k_{eff}\lambda_t}$	$\dfrac{4.064\pi}{2hn_{hf}L_f\sqrt{k_{eff}\lambda_t}\sqrt{(c_t\varphi)_{f+m}}}$
II	$\dfrac{141.2 S_{hf}^{face}}{hn_{hf}k_{eff}\lambda_t}$	$\dfrac{4.064\pi\sqrt{t}}{2hn_{hf}L\sqrt{k_{eff}\lambda_t}\sqrt{(c_t\varphi)_{f+m}}}$
III	$\dfrac{141.2 S_{hf}^{face}}{hn_{hf}k_{eff}\lambda_t}$	$\dfrac{4.064\pi\sqrt{t}}{2hn_{hf}L\sqrt{k_{eff}\lambda_t}\sqrt{(c_t\varphi)_{f+m}}}$

By taking advantage of straight-line analyses in Table 4.7, parameters like effective formation permeability and frac/HW half-length from the slope and skin factor from the intercept can be calculated in the case of linear flow.

4.2.2.3 Type Curve Matching Analysis

Using TC matching method in RTA of URs is very similar to primary RTA by Fetkovich, Blasingame, Agarwal-Gardner, and NPI TCs, however, in these methods, flow and fluid type assumptions in URs are not taken into consideration and therefore newer TCs to interpret production data of such reservoirs are required. This being said, occasionally, primary RTA methods have been used to analyze shale and tight reservoirs' production data. Though, the most accurate type of TC matching in URs' RTA would be following the basic flow and fluid type assumptions of unconventional plays.

As it was explained in Chap. 3, for TC matching of PTA in shale gas reservoirs (SGRs), Eq. 3.43 represents dimensionless pseudo-pressure which should be considered for each grid of HFs when the principle of superposition on different frac stages in a MFHW (according to Fig. 3.14) is being applied. The magnitude of this dimensionless pseudo-pressure in Laplace space is [29]:

$$\overline{m}_{Di}(P_f) = \frac{q_{ri}}{2L_{fDi}}\frac{1}{s}\int_{-L_{fDi}}^{+L_{fDi}} K_0\left[\sqrt{sf(s)}\sqrt{(x_D - x_{Di} - \xi)^2 + (y_D - y_{Di})^2}\right]d\xi \quad (4.36)$$

Now, using the principle proposed by van Everdingen and Hurst (1949) [27] if we aim to achieve the production flowrate response at constant BHFP conditions from pressure response at constant wellbore flowrate conditions in Laplace space, Eq. 4.37 can be introduced. Through this equation, dimensionless flowrate response of the SGR model in Laplace space ($\overline{q}_D$) is expressed in terms of dimensionless pseudo-pressure in Laplace space ($\overline{m}_D(P)$) as:

$$\overline{q}_D = \frac{1}{s^2 \overline{m}_D(P)} \quad (4.37)$$

It should be noticed that alike primary RTA methods, dimensionless flowrate would turn equal to inverse of the dimensionless pseudo-pressure:

$$q_D = \frac{1}{m_D(P)} = \frac{50,300 P_{sc} T}{k_f h T_{sc}} \frac{q_{sc}}{\Delta m(P)} \qquad (4.38)$$

where dimensionless pseudo-pressure is the same as the one in Table 3.2.

Ultimately, by taking advantage of Stehfest [25] numerical Laplace inversion (described in Appendix A) and following the procedure that was used in Chap. 3 in such well-SGR reservoir model for Eq. 4.37 (instead of Eq. 3.43), one can plot the TC of the RTA. MATLAB codes for creating these diagrams are described in Appendix C. Figures 4.4, 4.5, and 4.6 demonstrates the RTA TC plots of the Zhao et al. SGR model with sensitivity to Langmuir volume (V_L), HW length ($2L$), and number of HFs (n_{hf}), respectively.

Figure 4.4 exhibits that the amount of Langmuir volume, affects the inter-porosity and radial flows. As the amount of Langmuir volume increases, the dimensionless flowrate would increase and more significant reduction in the rate derivative curve should be expected. According to Figs. 4.5 and 4.6, based on the PTA that was presented in Chap. 3, the length of HW would affect the secondary linear and elliptical flows and the number of HFs would impact the primary linear and elliptical flows. It is also concluded that larger HW length, as well as when more fracs are created crossing the well, will result in a larger dimensionless flowrate.

Type curve matching for the purpose of RTA in this model can be done by using the definition of dimensionless parameters (introduced in Table 3.2 and Eq. 4.38) and selecting a proper matching point. Then, desired parameters can be calculated by comparing dimensional and dimensionless parameters as described earlier in Chap. 3. In RTA applications, after selecting a matching point on the matched curve of TC

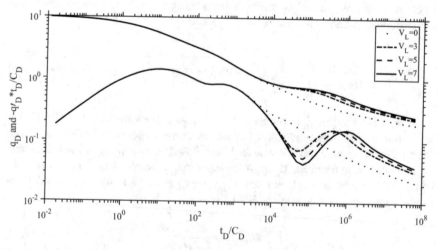

Fig. 4.4 RTA type curve plot of MFHW in the SGR with sensitivity to Langmuir volume

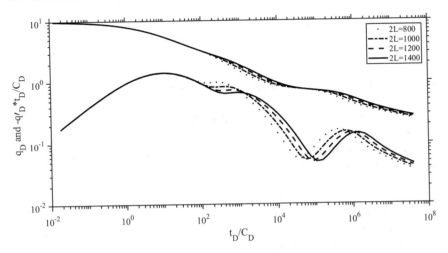

Fig. 4.5 RTA type curve plot of MFHW in the SGR with sensitivity to well length (2L)

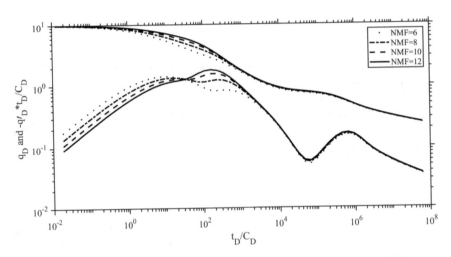

Fig. 4.6 RTA type curve plot of MFHW in the SGR with sensitivity to the number of HFs

and production data, one can calculate the HW half-length and natural fracture (NF) permeability using Eqs. 4.39 and 4.40, respectively.

$$L = \sqrt{\frac{0.006328 k_f}{(\varphi_f c_{tf} + \varphi_m c_{tm})\mu} \left(\frac{t}{t_D}\right)_M} \tag{4.39}$$

$$k_f = \frac{50,300 P_{sc} T}{h T_{sc}} \left(\frac{\frac{q_{sc}}{\Delta m(P)}}{q_D}\right)_M \tag{4.40}$$

In these formulations the terms with subscript "M" refers to the parameters from the matching point.

4.2.2.4 Linear Flow RTA in Rectangular Reservoirs

Linear flow regime in reservoirs with hydraulically fractured wells is a common phenomenon and the study of this type of flow geometry—especially in long term—is essential. It should be noted that in production from tight and shale reservoirs, linear flow lasts for several years [19] which points to the importance of analyzing this type of flow in fractured wells in URs. Apart from typical studies of flow in reservoirs with circular discharge area, due to the lateral extent of URs and the impact of rectangular drainage areas in them, the emphasis of this section is on the linear flow in such reservoirs. In the following section, linear flow equation in hydraulically fractured wells, and RTA in homogenous and naturally fractured URs with linear flow are discussed.

Linear Flow Equation in Wells with Vertical Fracs

Here we assume a reservoir with rectangular drainage area with $2x_e$ and $2y_e$ dimensions that contain a well with an ICHF created along the thickness of the reservoir (h) in the center and producing with constant rate q. In this system, HF half-length (L_f) is equal to x_e and the distance of the well from the reservoir boundary is y_w, as depicted in Fig. 4.7. Therefore, linear flow equation, when fluid propagate to the HF in this well-reservoir system can be defined by Eq. 4.41.

$$\frac{\partial^2 P}{\partial y^2} = \frac{\mu c_t \varphi}{0.00633k} \frac{\partial P}{\partial t} \tag{4.41}$$

Assuming the pressure is equal to the initial pressure of the reservoir at time t_0, the production flowrate at the wellbore would be constant, and there will not be any flow at the reservoir boundary; the solutions for initial, inner boundary, and outer boundary conditions of above equation are Eqs. 4.42, 4.43, and 4.44, respectively.

$$P(y, 0) = P_i \tag{4.42}$$

$$\left. \frac{\partial P}{\partial y} \right|_{y=0} = \frac{443.6q\mu B}{L_f k h} \tag{4.43}$$

$$\left. \frac{\partial P}{\partial y} \right|_{y=y_e} = 0 \tag{4.44}$$

Fig. 4.7 Top view schematic of linear flow in a rectangular reservoir having a well with vertical frac (Modified from [26])

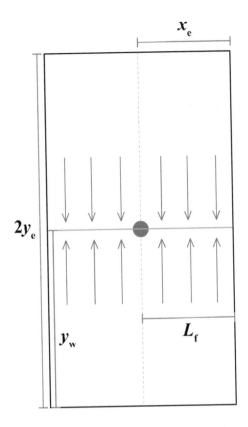

Based on the solution by Miller (1962) [18] for this PDE, the pressure response is in the form of Eq. 4.45.

$$P(y, t) = P_i -$$

$$\frac{443.6q\mu B}{L_f k h} \left[\left(\frac{(y_e - y)^2}{2y_e} - \frac{y_e}{2} + \frac{\eta t}{y_e} \right) - \frac{2y_e}{\pi^2} \sum_{n=1}^{\infty} \left(\frac{\exp\left(-\eta\left(\frac{n\pi}{y_e}\right)^2 t\right)}{n^2} \cos\left(\frac{n\pi y}{y_e}\right) \right) \right]$$

$$(4.45)$$

where:

$$\eta = \frac{0.00633k}{\mu c_t \varphi} \qquad (4.46)$$

By considering $y = 0$ to consider the location of the well, one can have:

$$P(0,t) = P_i - \frac{443.6q\mu B}{L_f kh}\left[\left(\frac{y_e}{3} + \frac{\eta t}{y_e}\right) - \frac{2y_e}{\pi^2}\sum_{n=1}^{\infty}\frac{\exp\left(-\eta\left(\frac{n\pi}{y_e}\right)^2 t\right)}{n^2}\right] \quad (4.47)$$

By defining dimensionless pressure and time, respectively, Eqs. 4.48 and 4.49, the magnitude of dimensionless pressure of the well is obtained by Eq. 4.50.

$$P_D = \frac{kh(P_i - P)}{141.2q\mu B} \quad (4.48)$$

$$t_{DL_f} = \frac{0.00633kt}{\mu c_t \varphi L_f^2} \quad (4.49)$$

$$P_D = \frac{\pi}{2}\left[\left(\frac{y_e}{3L_f} + \frac{L_f t_{DL_f}}{y_e}\right) - \frac{2y_e}{\pi^2 L_f}\sum_{n=1}^{\infty}\frac{\exp\left(-\left(\frac{n\pi L_f}{y_e}\right)^2 t_{DL_f}\right)}{n^2}\right] \quad (4.50)$$

Now, by defining the dimensionless pressure and time of decline, respectively, in the form of Eqs. 4.51 and 4.52, the dimensionless pressure of decline results in Eq. 4.53.

$$P_{Dd} = \frac{L_f}{y_e}P_D \quad (4.51)$$

$$t_{Dd} = \left(\frac{L_f}{y_e}\right)^2 t_{DL_f} \quad (4.52)$$

$$P_{Dd} = \frac{\pi}{2}\left[\left(\frac{1}{3} + t_{Dd}\right) - \frac{2}{\pi^2}\sum_{n=1}^{\infty}\frac{\exp\left(-(n\pi)^2 t_{Dd}\right)}{n^2}\right] \quad (4.53)$$

Also, production flowrate solution of the flow equation under constant BHFP condition can be worked using the principle introduced by [27], as illustrated in Eq. 4.37, while the inverse of dimensionless decline flowrate is presented as:

$$\frac{1}{q_{Dd}} = \frac{\pi}{\sum_{n=1,3,5,\dots}^{\infty}\exp\left(-\left(\frac{n\pi}{2}\right)^2 t_{Dd}\right)} \quad (4.54)$$

Linear Flow RTA in Homogenous URs

Following the method developed by Wattenbarger et al. (1998) [28], approximate solution of the linear flow in rectangular reservoir for early-/late-times are as summarized in Table 4.8.

Table 4.8 Dimensionless pressure/rate approximate solutions for early-/late-times of linear flow rectangular reservoir

Solution	Formulation
Early-time pressure solution	$P_{wD} = \sqrt{\pi t_{DL_f}}$
Early-time rate solution	$\frac{1}{q_D} = \frac{\pi}{2}\sqrt{\pi t_{DL_f}}$
Late-time pressure solution	$P_{wD} = \frac{\pi}{2}\left(\frac{L_f}{y_e}\right)t_{DL_f} + \frac{\pi}{6}\left(\frac{y_e}{L_f}\right)$
Late-time rate solution	$\frac{1}{q_D} = \frac{\pi}{4}\left(\frac{y_e}{L_f}\right)\exp\left(\frac{\pi^2}{4}\left(\frac{L_f}{y_e}\right)^2 t_{DL_f}\right)$

Table 4.9 Dimensionless decline pressure/rate approximate solutions for early-/late-times of linear flow rectangular reservoir

Solution	Formulation
Early-time pressure solution	$P_{Dd} = \sqrt{\pi t_{Dd}}$
Early-time rate solution	$\frac{1}{q_{Dd}} = \frac{\pi}{2}\sqrt{\pi t_{Dd}}$
Late-time pressure solution	$P_{Dd} = \frac{\pi}{2}t_{Dd} + \frac{\pi}{6}$
Late-time rate solution	$\frac{1}{q_{Dd}} = \frac{\pi}{4}\exp\left(\frac{\pi^2}{4}t_{Dd}\right)$

By defining dimensionless decline pressure, rate inverse, and time as Eqs. 4.55, 4.56, and 4.57, respectively, dimensionless decline solution of early-/late-times are similar to those shown in Table 4.9.

$$P_{Dd} = \frac{L_f}{y_e}P_{wD} \tag{4.55}$$

$$\frac{1}{q_{Dd}} = \frac{L_f}{y_e}\frac{1}{q_D} \tag{4.56}$$

$$t_{Dd} = \left(\frac{L_f}{y_e}\right)^2 t_{DL_f} = \frac{kt}{\mu c_t \varphi y_e^2} = t_{Dy_e} \tag{4.57}$$

In Fig. 4.8, dimensionless decline pressure curve (at constant flowrate conditions) and dimensionless decline inverse rate curve (at constant BHFP conditions) in terms of dimensionless decline time ($t_{Dd} = t_{Dye}$), in both early—and late-times are plotted.

From the log–log plot of Fig. 4.8, it is seen that according to relationships presented in Table 4.9, the slope of the curves in the early-times—infinite acting linear flow (IALF)—would be equal to 0.5. Moreover, at later-times of the linear flow, based on P_{Dd} formulation, the unit slope of the dimensionless decline pressure diagram can be observed. In addition, considering $1/q_{Dd}$ formulation, the exponential changes of the inverse of the dimensionless decline rate can be found. Also, as it can be inferred from this figure, approximate ending of IALF is $t_{Dd} = 0.5$ for constant flowrate condition and $t_{Dd} = 0.25$ for constant BHFP condition.

TCs of constant rate (dimensionless pressure) and constant BHFP assumption (dimensionless rate inverse), based on the formulations laid out earlier, are plotted

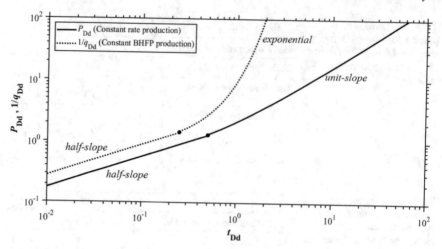

Fig. 4.8 Dimensionless pressure and rate inverse response of linear flow in rectangular reservoir

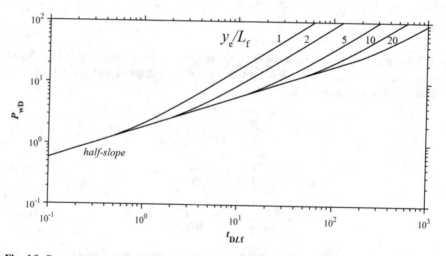

Fig. 4.9 Pressure type curves of constant rate production from rectangular reservoir

in Figs. 4.9 and 4.10, respectively. The following plots are known as Wattenbarger type curves in RTA literature.

Using these TCs, similar procedure is commonly perused in all TC matching workflows, by matching the field data with one of the TCs, based on dimensionless rate/pressure/time relation with their corresponding dimensional values, enabling us to find unknown properties of the well/reservoir. In the following, formulations necessary for calculating drainage radius (A_e), hydrocarbon in-place (N for oil reservoir and G for gas reservoir) and some other terms such as: well-reservoir properties, in a TC matching for linear flow in a rectangular reservoir, are presented for oil

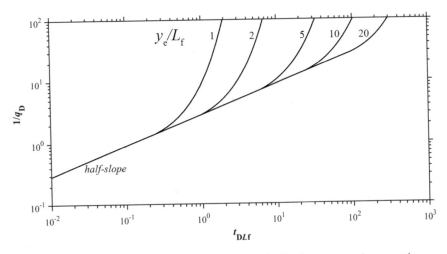

Fig. 4.10 Rate inverse type curves of constant BHFP production from rectangular reservoir

(Eqs. 4.58, 4.59, 4.60, 4.61, and 4.62) and gas (Eqs. 4.63, 4.64, 4.65, 4.66, and 4.67) reservoirs. Please note, here, pseudo-time (for gas reservoirs) and MBT concepts are applied in conjunction with constant BHFP production relationships to smooth the data.

Formulations for property calculation in TC matching of RTA linear flow of rectangular oil reservoir are as follows:

$$\frac{kL_f}{y_e} = \frac{141.2\mu B}{h} \frac{\left(\frac{q}{\Delta P}\right)_M}{(q_{Dd})_M} \tag{4.58}$$

$$\frac{y_e}{\sqrt{k}} = \sqrt{\frac{0.00633}{\mu c_t \varphi}\left(\frac{t_c}{t_{cDd}}\right)_M} \tag{4.59}$$

$$\sqrt{k}L_f = \frac{141.2\mu B}{h}\frac{\left(\frac{q}{\Delta P}\right)_M}{(q_{Dd})_M}\sqrt{\frac{0.00633}{\mu c_t \varphi}\left(\frac{t_c}{t_{cDd}}\right)_M} \tag{4.60}$$

$$A_e = 4x_e y_e = 4\sqrt{k}L_f\frac{y_e}{\sqrt{k}} = \frac{0.8938B}{hc_t\varphi}\frac{\left(\frac{q}{\Delta P}\right)_M}{(q_{Dd})_M}\left(\frac{t_c}{t_{cDd}}\right)_M \tag{4.61}$$

$$N = \frac{7758.36A_e h\varphi S_o}{B_o} \tag{4.62}$$

Similarly, other formulations to calculate reservoir properties in a linear flow of rectangular gas reservoir are:

$$\frac{kL_f}{y_e} = \frac{1424\mu B \left(\frac{q}{\Delta m(P)}\right)_M}{h} \frac{1}{(q_{Dd})_M} \tag{4.63}$$

$$\frac{y_e}{\sqrt{k}} = \sqrt{\frac{0.00633}{\mu c_t \varphi} \left(\frac{t_{ca}}{t_{caDd}}\right)_M} \tag{4.64}$$

$$\sqrt{k}L_f = \frac{1424\mu B \left(\frac{q}{\Delta m(P)}\right)_M}{h} \frac{1}{(q_{Dd})_M} \sqrt{\frac{0.00633}{\mu c_t \varphi} \left(\frac{t_{ca}}{t_{caDd}}\right)_M} \tag{4.65}$$

$$A_e = 4x_e y_e = 4\sqrt{k}L_f \frac{y_e}{\sqrt{k}} = \frac{9.0139B \left(\frac{q}{\Delta m(P)}\right)_M}{hc_t \varphi} \frac{1}{(q_{Dd})_M} \left(\frac{t_{ca}}{t_{caDd}}\right)_M \tag{4.66}$$

$$G = \frac{43560 A_e h\varphi S_g}{B_{gi}} \tag{4.67}$$

where B_{gi} denotes the formation volume factor of gas at initial reservoir pressure.

Linear Flow RTA in Naturally Fractured URs

Considering the matrix of the reservoir shown in Fig. 4.7 to be naturally fractured, one can use dual-porosity model to develop required RTA relationships under this assumption for such rectangular reservoirs.

Governing flow equations for naturally fractured rectangular URs producing with vertical HF is the following expression [8]:

$$\frac{\partial^2 P_{Df}}{\partial y_D^2} = \omega \frac{\partial P_{Df}}{\partial t_{DA_c}} + (1-\omega)\frac{\partial P_{Dm}}{\partial t_{DA_c}} \tag{4.68}$$

where subscripts "f" and "m" refer to fracture and matrix, respectively. Also, A_c is the reservoir cross-sectional area, $A_c = 4L_f h$, and ω is the fracture storativity ratio that was introduced earlier (ω_f).

Considering the state of flow regime in the matrix medium, flow equation in matrix with PSS and transient flows are Eqs. 4.69 and 4.70, respectively.

$$\frac{\partial P_{Dm}}{\partial t_{DA_c}} = \frac{\lambda}{1-\omega}(P_{Df} - P_{Dm}) \tag{4.69}$$

$$\frac{\partial^2 P_{Dm}}{\partial y_D^2} = \frac{3(1-\omega)}{\lambda}\frac{\partial P_{Df}}{\partial t_{DA_c}} \tag{4.70}$$

Table 4.10 Initial and boundary conditions of flow equation in naturally fractured rectangular reservoir producing with vertical ICHF

IC/BC	Formulation
IC at NF	$P_{Df}(y_D, 0) = 0$
IC at matrix	PSS matrix flow: $P_{Dm}(y_D, 0) = 0$
	Transient matrix flow: $P_{Dm}(z_D, 0) =$
	$0, \left. \frac{\partial P_{Dm}}{\partial z_D} \right|_{z_D=0} = 0$
Inner BC (constant rate production)	$\left. \frac{\partial P_{Df}}{\partial y_D} \right|_{y_D=0} = -2\pi$
Inner BC (constant BHFP production)	$P_{Df}(y_D = 0, t_{DA_c}) = 1$
Outer BC (infinite reservoir)	$P_{Df}(y_D \to \infty, t_{DA_c}) = 0$
Outer BC (closed reservoir)	$\left. \frac{\partial P_{Df}}{\partial y_D} \right|_{y_D = \frac{2y_{eD}}{A_c}} = 0$
Outer BC (constant pressure boundary)	$P_{Df}\left(y_D = \frac{2y_{eD}}{A_c}, t_{DA_c}\right) = 0$
Matrix-fracture flow BC (transient matrix case)	$P_{Dm}(1, t_{DA_c}) = P_{Df}(1, t_{DA_c})$

Table 4.11 El-Banbi dimensionless parameters

Dimensionless parameter	Formulation
Dimensionless distance in y-direction	$y_D = \frac{y}{\sqrt{A_c}}$
Dimensionless time	$t_{DA_c} = \frac{0.00633k_f t}{\mu(c_{tf}\varphi_f + c_{tm}\varphi_m)A_c}$
Dimensionless pressure (Constant rate production)	$P_D = \frac{k_f\sqrt{A_c}(P_i - P_{wf})}{141.2q\mu B}$
Dimensionless pressure (Constant BHFP production)	$P_D = \frac{P_i - P}{P_i - P_{wf}}$
Dimensionless rate inverse (Constant BHFP production)	$\frac{1}{q_D} = \frac{k_f\sqrt{A_c}(P_i - P_{wf})}{141.2q\mu B}$

Initial and boundary conditions (ICs and BCs) of flow equation, considering both PSS and transient matrix flows, are as presented in Table 4.10.

Also, El-Banbi dimensionless parameters used in his method are summarized in Table 4.11.

Based on [8] workflow, final Laplace space pressure solution of flow equation in naturally fractured rectangular oil reservoirs producing with vertical fracs considering infinite-acting, closed, and constant pressure boundaries of the reservoir would become Eqs. 4.71, 4.72, and 4.73, respectively.

$$\overline{P}_{wD} = \frac{2\pi}{s\sqrt{sf(s)}} \tag{4.71}$$

$$\overline{P}_{wD} = \frac{2\pi}{s\sqrt{sf(s)}}\left[\frac{1 + \exp\left(-2\sqrt{sf(s)}\frac{2y_{eD}}{\sqrt{A_c}}\right)}{1 - \exp\left(-2\sqrt{sf(s)}\frac{2y_{eD}}{\sqrt{A_c}}\right)}\right] \tag{4.72}$$

$$\overline{P}_{wD} = \frac{2\pi}{s\sqrt{sf(s)}}\left[\frac{1 - \exp\left(-2\sqrt{sf(s)}\frac{2y_{eD}}{\sqrt{A_c}}\right)}{1 + \exp\left(-2\sqrt{sf(s)}\frac{2y_{eD}}{\sqrt{A_c}}\right)}\right] \tag{4.73}$$

In addition to solution that was derived based on constant rate production assumption, dimensionless rate inverse solution in Laplace space for the same reservoir considering infinite-acting, closed, and constant pressure boundaries would be Eqs. 4.74, 4.75, 4.76, respectively.

$$\frac{1}{q_D} = \frac{2\pi s}{\sqrt{sf(s)}} \tag{4.74}$$

$$\frac{1}{q_D} = \frac{2\pi s}{\sqrt{sf(s)}}\left[\frac{1 + \exp\left(-2\sqrt{sf(s)}\frac{2y_{eD}}{\sqrt{A_c}}\right)}{1 - \exp\left(-2\sqrt{sf(s)}\frac{2y_{eD}}{\sqrt{A_c}}\right)}\right] \tag{4.75}$$

$$\frac{1}{q_D} = \frac{2\pi s}{\sqrt{sf(s)}}\left[\frac{1 - \exp\left(-2\sqrt{sf(s)}\frac{2y_{eD}}{\sqrt{A_c}}\right)}{1 + \exp\left(-2\sqrt{sf(s)}\frac{2y_{eD}}{\sqrt{A_c}}\right)}\right] \tag{4.76}$$

In these Laplace space solutions, the term $f(s)$ is related to inter-porosity flow and has different mathematical expressions for each case of PSS and transient matrix flows. As a matter of fact, $f(s)$ term takes the NFs terms storativity ratio (ω) and inter-porosity flow coefficient (λ) into account for pressure/rate solutions of flow in the reservoir. Table 4.12 presents different $f(s)$ terms for homogenous reservoir, and naturally fractured rectangular UOR producing with vertical ICHF considering PSS

Table 4.12 Different $f(s)$ terms for dual-porosity rectangular reservoir producing vertical frac

Reservoir model	$f(s)$ term
Homogenous	$f(s) = 1$
PSS matrix flow dual-porosity	$f(s) = \frac{\omega(1-\omega)s+\lambda}{(1-\omega)s+\lambda}$
Transient matrix dual-porosity (cubes model of matrix blocks)	$f(s) = \omega + \frac{\lambda}{5s}\sqrt{\frac{15(1-\omega)s}{\lambda}}\,coth\left(\sqrt{\frac{15(1-\omega)s}{\lambda}}\right)$
Transient matrix dual-porosity (sticks model of matrix blocks)	$f(s) = \omega + \frac{\lambda}{4s}\sqrt{\frac{8(1-\omega)s}{\lambda}}\,\dfrac{I_1\left(\sqrt{\frac{8(1-\omega)s}{\lambda}}\right)}{I_0\left(\sqrt{\frac{8(1-\omega)s}{\lambda}}\right)}$
Transient matrix dual-porosity (slabs model of matrix blocks)	$f(s) = \omega + \frac{\lambda}{3s}\sqrt{\frac{3(1-\omega)s}{\lambda}}\,tanh\left(\sqrt{\frac{3(1-\omega)s}{\lambda}}\right)$

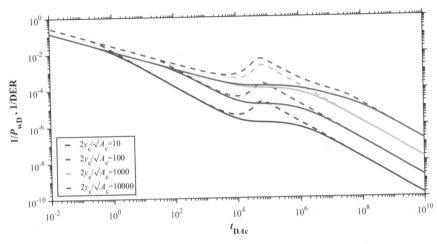

Fig. 4.11 Pressure inverse type curves of constant rate production from naturally fractured rectangular reservoir ($\omega = 0.01$, $\lambda = 0.000001$)

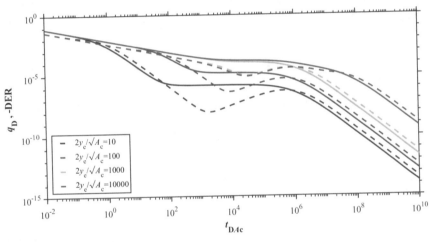

Fig. 4.12 Rate type curves of constant BHFP production from naturally fractured rectangular reservoir ($\omega = 0.01$, $\lambda = 0.000001$)

and transient flow assumptions in matrix medium. It should be emphasized that in a transient flow, three types of matrix blocks are considered; cubes, sticks, slabs.

Figures 4.11 and 4.12 depict El-Banbi TCs for RTA of naturally fractured rectangular reservoir producing with vertical fracs. Figure 4.11 is the inverse of pressure and Fig. 4.12 is rate which are related to constant rate and BHFP production assumptions, respectively. In these figures, in addition to pressure/rate (solid curves), the derivative of pressure/rate (dashed curves) is plotted too.

From these figures, we deduce that in the early times, the half-slope while the unit-slope in the later-times for either of the plots and also for both pressure/rate and their derivatives can be used. In the mid-times, inter-porosity flow occurs which is characterized with a bump on derivative curves. Alike all TC matching procedures, correlation between dimensionless parameters and their dimensional counterparts would enable us to calculate well/reservoir properties.

4.2.2.5 Modification of UGRs Linear Flow Analysis

It was explained previously that in linear flow regime, $\sqrt{k}L_f$ can be calculated from the slope of the line when normalized pressure-drop ($\Delta P/q$) versus square root of time ($\sqrt{t}$) is plotted via production from hydraulically fractured wells in UORs/UGRs. Researchers [12, 13, 19, 21] have found that $\sqrt{k}L_f$ which is estimated following this method in UGR wells is technically overestimated, despite the use of pseudo-pressure and pseudo-time concepts. In order to amend this error and remove the overestimation in the slope of the line, a correction factor (f_{cp}) has been introduced. Therefore, multiplication of $\sqrt{k}L_f$ by the corretion factor, will fix the overestimation (slope of the line). The experimental correction factor, f_{cp} is defined by Eq. 4.77 [12].

$$f_{cp} = 1 - 0.0852D_D - 0.0857D_D^2 \qquad (4.77)$$

where D_D is the draw-down parameter and is defined by the following equation:

$$D_D = \frac{m(P_i) - m(P_{wf})}{m(P_i)} \qquad (4.78)$$

Besides, the magnitude of correction factor, f_{cp}, has been proven analytically [19], defined as:

$$f_{cp} = \sqrt{\frac{\mu_{gi}c_{ti}}{\overline{\mu}_g\overline{c}_t}} \qquad (4.79)$$

where μ_{gi} and c_{ti}, are viscosity and compressibility of the gas in the initial conditions, respectively, and $\overline{\mu}_g$ and $\overline{c}_t$ denote the viscosity and compressibility of the gas at average reservoir pressure, respectively.

Likewise, another deviation from the true-behavior of standard straight-line trend in the analysis of UGRs happens in production data in those plots, occurring during the linear flow regime period. In this case, although linear flow regime is dominant in that period, a straight-line trend cannot be observed in the production analysis plots.

To address this issue, Nobakht and Clarkson [20] has defined the modified pseudo-time (t_a^*) according to Eq. 4.80.

$$t_a^* = t - \frac{q B_{gi} \sqrt{\mu_{gi} c_{gi} \varphi}}{6 h \varphi S_{gi} L_f \sqrt{k}} t \sqrt{t} \qquad (4.80)$$

In the early-times, the value of t_a^* is equal to t, but in the late-times it becomes greater than t; $t_a^* > t$. Furthermore, by replacing the value of t_a^* with t in late-times, the error in plotting $\Delta m(P)$ versus $\sqrt{t}$ can be fixed, and the straight-line analysis—with the same ideal slope— is delineated if $\Delta m(P)$ versus $\sqrt{t_a^*}$ is plotted.

4.2.2.6 Dynamic Slippage Effect in UGRs

To be more practical when analyzing RTA data in UGRs', especially shale and tight gas reservoirs, concepts specific to such reservoirs such as gas sorption and gas rarefaction (dynamic slippage) should be considered in the basic equations of gas flow, in a practical and applicable way.

For this purpose, by considering the Darcian flow to include the pressure-driven flow and the Fickian diffusion to include the concentration-driven flow, the gas permeability is defined as follows:

$$k_g = \left(1 + \frac{b_a}{P}\right)^{-1} \qquad (4.81)$$

In this equation, P is the pressure of the system and b_a is the gas slippage coefficient, while following [9], gas slippage can be defined as:

$$b_a = \frac{P c_g \mu_g D}{0.00633 k_\infty} \qquad (4.82)$$

where D is the diffusion coefficient (ft^2/day) and k_∞ is the equivalent liquid permeability (md).

To consider the effects of gas sorption and apparent gas permeability in the RTA relationships of UGRs, modified pseudo-pressure and pseudo-time values are defined respectively. Based on the method proposed by Clarkson et al. [7], modified pseudo-pressure and pseudo-time expressions are introduced in the form of Eqs. 4.83 and 4.84, respectively.

$$m^*(P_i) - m^*(P_{wf}) = 2 \int_{P_{wf}}^{P_i} \frac{1}{k_g(P) \mu_g Z} P \, dP \qquad (4.83)$$

$$t_a^* = \mu_{gi} c_{ti}^* \int_0^t \frac{dt}{\overline{k_g \mu_g c_t^*}} \qquad (4.84)$$

where c_t^* is the total isothermal compressibility coefficient in conjunction with considering the compressibility of the adsorbed gas as one of the production phases. Moreover, parameters that have "i" and "—" in sub/superscription, represent initial and average reservoir pressure, respectively.

A variety of straight-line plots for bilinear, linear, elliptical, pseudo-radial, and boundary-dominated flow regimes RTA, using modified pseudo-pressure and pseudo-time definitions are compiled in 4.12 according to the study by Clarkson and Beierle [5]. It should be noted that interpretation (plotting data, characteristics in logarithmic plot, and obtained properties) in this table can work for both modified/regular definitions of pseudo-functions for UGRs with/without considering dynamic slippage, respectively.

In Table 4.13, q_g is gas production rate (MSCF/day G_i is initial gas in-lace (MMSCF), w_f fracture width (ft). Other parameters are in field unit, while, m and b', are slope and y-intercept of the corresponding plots, respectively.

Radial, linear, and bilinear derivatives mentioned in 4.12, are obtained by taking the derivative from normalized pseudo-pressure with respect to $\ln(t)$, $t^{0.25}$, and $t^{0.5}$, respectively. Similarly, in the straight-line analyses related to the bilinear, linear, and pseudo-radial flow regimes, in addition to plotting normalized pseudo-pressure data versus $\left(t_a^*\right)^{\frac{1}{4}}, \left(t_a^*\right)^{\frac{1}{2}}$, and $\log\left(t_a^*\right)$, respectively, they can be plotted versus $\sum\limits_{j=1}^{n} \frac{q_j - q_{j-1}}{q_n}\left(t_{a,n}^* - t_{a,j-1}^*\right)^{\frac{1}{4}}, \sum\limits_{j=1}^{n} \frac{q_j - q_{j-1}}{q_n}\left(t_{a,n}^* - t_{a,j-1}^*\right)^{\frac{1}{2}}$, and $\sum\limits_{j=1}^{n} \frac{q_j - q_{j-1}}{q_n} \log\left(t_{a,n}^* - t_{a,j-1}^*\right)$, correspondingly if flow-rates in the production history is changing. These expressions, known as superposition time, would better take into account the effect of n-times change in production rate, while analyzing the production data.

For an elliptical flow, parameters A and B in ICHF case, and parameters A' and B' in FCHF case should be determined prior to carrying the analysis out. These parameters are related to one another via:

$$A = \sqrt{B^2 + L_f^2} \tag{4.85}$$

$$A' = \sqrt{B'^2 + L_f^2} \tag{4.86}$$

Parameter B for the early-times of elliptical flow and also later-times when elliptical flow is developed to frac's half-lengths can be measured using Eqs. 4.87 and 4.88, respectively [11, 14].

$$B = 0.02878\sqrt{\frac{kt}{\mu c_t \varphi}} \tag{4.87}$$

Table 4.13 Summary of straight-line RTA plots of different flow regimes in SGR considering dynamic slippage and producing with hydraulically fractured well

Properties obtained from linear analysis	Characteristic in logarithmic plot	Linear analysis diagram (constant flow)	Flow regime
$\sqrt{w_f k_{hf}}(k)^{\frac{1}{4}} = \dfrac{443.2T}{mh(\mu_{gi}c_{ti}^*\varphi)^{\frac{1}{4}}}$	Radial derivative: 0.25 slope; Bilinear derivative: zero slope	$\dfrac{m^*(P_i)-m^*(P_{wf})}{q_g}$ versus $(t_a^*)^{\frac{1}{4}}$ or $\sum_{j=1}^{n}\dfrac{q_j-q_{j-1}}{q_n}\left(t_{a,n}^* - t_{a,j-1}^*\right)^{\frac{1}{4}}$	Bilinear
$L_f\sqrt{k} = \dfrac{40.93T}{mh(\mu_{gi}c_{ti}\varphi)^{\frac{1}{2}}}$	Radial derivative: 0.5 slope; Linear derivative: zero slope	$\dfrac{m^*(P_i)-m^*(P_{wf})}{q_g}$ versus $(t_a^*)^{\frac{1}{2}}$ or $\sum_{j=1}^{n}\dfrac{q_j-q_{j-1}}{q_n}\left(t_{a,n}^* - t_{a,j-1}^*\right)^{\frac{1}{2}}$	Linear
$k = \dfrac{1422T}{mh}$ $L_f = \exp(b')$	Radial Derivative: Zero Slope	$\dfrac{m^*(P_i)-m^*(P_{wf})}{q_g}$ versus $\ln(A+B)$ (ICHF) or $\ln(A'+B')$ (FCHF)	Elliptical
$k = \dfrac{1637T}{mh}$ $S = 1.1513\left[\dfrac{b'}{m} - \log\left(\dfrac{k}{\mu_{gi}c_{ti}^*\varphi r_w^2}\right) + 3.23\right]$	Radial Derivative: zero slope	$\dfrac{m^*(P_i)-m^*(P_{wf})}{q_g}$ versus $\log(t_a^*)$ or $\sum_{j=1}^{n}\dfrac{q_j-q_{j-1}}{q_n}\log\left(t_{a,n}^* - t_{a,j-1}^*\right)$	Pseudo-radial
Initial gas in-place (G_i) of x-intercept	Radial derivative: unit slope	$\dfrac{q_g}{m^*(P_i)-m^*(P_{wf})}$ versus $G_i\dfrac{m^*(P_i)-m^*(P_R)}{m^*(P_i)-m^*(P_{wf})}$	BDF

$$B = 0.02634 \sqrt{\frac{kt}{\mu c_t \varphi}} \tag{4.88}$$

Also B' is related to B; $B' = R \times B$. Where parameter R is expressed as [4]:

$$R = \frac{L_f}{L_{fe}} = \frac{\pi}{2F_{CD}} + 1 \tag{4.89}$$

where F_{CD} is dimensionless FCHF conductivity defined as Eq. 3.53 in Chap. 3.

Another important point in straight-line analyses of Table 4.12 is to ensure that the time period when production rate is being analyzed, should be truly consistent with the corresponding time period when desired flow regime is expected to occur in the vicinity of the wellbore. To confirm this consistency in the data and predefined flow regime models, as it is outlined in Table 4.12, one can use the characteristics of derivative curves plotted in logarithmic scale. Finally, since fracs exist in both vertical and horizontal wells, also they can be single or multiple in the order of occurrence in the wells, their impact on flow regimes might not be similar. In the following, order of occurrence for different flow regimes; (1) vertical well with single frac, (2) horizontal well without any frac, and (3) horizontal well with multiple fracs (MFHW) are shown schematically on derivative plots of modified normalized pseudo-pressure data versus modified pseudo-time, and also summarized separately in tables, in Figs. 4.13, 4.14, and 4.15, respectively.

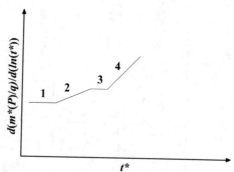

Order	Flow regime	Characteristic on logarithmic plot
1	Initial radial	Radial derivative: zero slope
2	Linear	Radial derivative: 0.5 slope
		Linear derivative: zero slope
3	Pseudo-radial	Radial derivative: zero slope
4	Boundary-dominated	Radial derivative: unit slope

Fig. 4.13 Graphical illustration of different flow regimes on RTA logarithmic plot derivative curve for a vertical well with a single frac (Modified from [6])

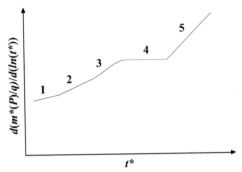

Order	Flow regime	Characteristic on logarithmic plot
1	Bilinear	Radial derivative: 0.25 slope
		Bilinear derivative: zero slope
2	Linear	Radial derivative: 0.5 slope
		Linear derivative: zero slope
3	Elliptical	Radial derivative: non-linear
4	Pseudo-radial	Radial derivative: zero slope
5	Boundary-dominated	Radial derivative: unit slope

Fig. 4.14 Graphical illustration of different flow regimes on RTA logarithmic plot derivative curve for a horizontal well with no frac (Modified from [6])

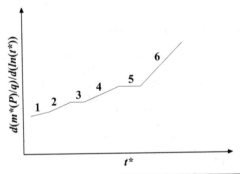

Order	Flow regime	Characteristic on logarithmic plot
1	Bilinear	Radial derivative: 0.25 slope
		Bilinear derivative: zero slope
2	Initial linear	Radial derivative: 0.5 slope
		Linear derivative: zero slope
3	Initial radial	Radial derivative: zero slope
4	Secondary linear	Radial derivative: 0.5 slope
		Linear derivative: zero slope
5	Secondary radial	Radial derivative: zero slope
6	Boundary-dominated	Radial derivative: unit slope

Fig. 4.15 Graphical illustration of different flow regimes on RTA logarithmic plot derivative curve for a MFHW (Modified from [6])

References

1. Agarwal RG, Gardner DC, Kleinsteiber SW, Fussell DD (1999) Analyzing well production data using combined type curve and decline curve analysis concepts. SPE Reserv Eval Eng 2:478–486. https://doi.org/10.2118/57916-PA
2. Arps JJ (1945) Analysis of Decline Curves. Trans AIME 160:228–247. https://doi.org/10.2118/945228-G
3. Blasingame TA, Johnston JL, Lee WJ (1989) Type-Curve Analysis Using the Pressure Integral Method . SPE Calif. Reg. Meet.
4. Cheng Y, Lee WJ, McVay DA (2009) A new approach for reliable estimation of hydraulic fracture properties in tight gas wells. SPE Reserv Eval Eng 12:254–262. https://doi.org/10.2118/105767-PA
5. Clarkson CR, Beierle JJ (2011) Integration of microseismic and other post-fracture surveillance with production analysis: a tight gas study. J Nat Gas Sci Eng 3:382–401
6. Clarkson CR, Jordan CL, Ilk D, Blasingame TA (2009) Production Data Analysis of Fractured and Horizontal CBM Wells. In: SPE Eastern Regional Meeting. Society of Petroleum Engineers, Charleston, West Virginia, USA
7. Clarkson CR, Nobakht M, Kaviani D, Ertekin T (2012) Production Analysis of Tight-Gas and Shale-Gas Reservoirs Using the Dynamic-Slippage Concept. SPE J 17:230–242
8. El-Banbi AH (1998) Analysis of Tight Gas Well Performance. Texas A&M University
9. Ertekin T, King GA, Schwerer FC (1986) Dynamic gas slippage: a unique dual-mechanism approach to the flow of gas in tight formations. SPE Form Eval 1:43–52
10. Fetkovich MJ (1980) Decline Curve Analysis Using Type Curves. J Pet Technol 32:1065–1077. https://doi.org/10.2118/4629-PA
11. Hale BW, Evers JF (1981) Elliptical flow equations for vertically fractured gas wells. J Pet Technol 33:2–489
12. Ibrahim M, Wattenbarger RA (2006a) Rate dependence of transient linear flow in tight gas wells. J Can Pet Technol 45https://doi.org/10.2118/06-10-TN2
13. Ibrahim MH, Wattenbarger RA (2006b) Analysis of Rate Dependence in Transient Linear Flow in Tight Gas Wells. Society of Petroleum Engineers (SPE)
14. Jones LG (1963) Reservoir Reserve Tests. J Pet Technol 15:333–337
15. Kurtoglu B (2013) Integrated Reservoir Characterization and Modeling in Support of Enhanced Oil Recovery for Bakken. Colorado School of Mines
16. Kurtoglu B, Cox SA, Kazemi H (2011) Evaluation of Long-Term Performance of Oil Wells in Elm Coulee Field. In: Canadian Unconventional Resources Conference. Society of Petroleum Engineers, Calgary, Alberta, Canada
17. Mattar L, Anderson D (2005) Dynamic material balance (oil or gas-in-place without shut-ins). In: Canadian International Petroleum Conference (CIPC). Petroleum Society of Canada
18. Miller FG (1962) Theory of unsteady-state influx of water in linear reservoirs. J Inst Pet 48:365–379
19. Nobakht M, Clarkson CR (2011a) A New Analytical Method for Analyzing Production Data from Shale Gas Reservoirs Exhibiting Linear Flow: Constant Pressure Production . North Am. Unconv. Gas Conf. Exhib.
20. Nobakht M, Clarkson CR (2011b) A New Analytical Method for Analyzing Production Data from Shale Gas Reservoirs Exhibiting Linear Flow: Constant Rate Production . North Am. Unconv. Gas Conf. Exhib.
21. Nobakht M, Mattar L, Moghadam S, Anderson DM (2010) Simplified Yet Rigorous Forecasting of Tight/Shale Gas Production in Linear Flow. In: SPE Western Regional Meeting. Society of Petroleum Engineers, Anaheim, California
22. Palacio JC, Blasingame TA (1993) Decline-Curve Analysis Using Type Curves--Analysis of Gas Well Production Data. In: SPE Joint Rocky Mountain Regional and Low Permeability Reservoirs Symposium. Society of Petroleum Engineers, Denver, Colorado
23. Rezaee R (2015) Fundamentals of Gas Shale Reservoirs. Wiley Online Library

24. Sabet MA (1991) Well Test Analysis. Gulf Professional Publishing
25. Stehfest H (1970) Numerical Inversion of Laplace Transform. Commun ACM 13:47–49. https://doi.org/10.1145/361953.361969
26. Sun H (2015) Advanced production decline analysis and application. Gulf professional publishing
27. van Everdingen AF, Hurst W (1949) The Application of the Laplace Transformation to Flow Problems in Reservoirs. J Pet Technol 1:305–324. https://doi.org/10.2118/949305-G
28. Wattenbarger RA, El-Banbi AH, Villegas ME, Maggard JB (1998) Production Analysis of Linear Flow Into Fractured Tight Gas Wells. In: SPE Rocky Mountain Regional/Low-Permeability Reservoirs Symposium. SPE, Denver, Colorado
29. Zhao Y long, Zhang L hui, Zhao J zhou, et al (2013) "Triple porosity" modeling of transient well test and rate decline analysis for multi-fractured horizontal well in shale gas reservoirs. J Pet Sci Eng 110:253–262. https://doi.org/10.1016/j.petrol.2013.09.006https://doi.org/10.1016/j.petrol.2013.09.006

Appendix A

Governing flow equations for pressure/rate transient analysis (PTA/RTA) in Chaps. 3 and 4 of this book, respectively, were solved semi-analytically which means they were only solved analytically in Laplace domain. The solutions were mainly dimensionless pressure in Laplace space. To reach time domain solution of these flow equations a Laplace inverse is needed. Due to complexity of mentioned pressure solutions, analytical Laplace inversion is not viable and numerical Laplace inverse methods should be used instead.

In the following, the method presented by [1], which is widely used in the area of PTA/RTA modeling, is presented in detail as a numerical method of Laplace inversion. This method should be applied on Laplace space pressure/rate solutions to reach their corresponding time domain solutions. In this method, by considering Laplace space pressure term as $P(s)$, its Laplace inverse is as $f(t) = L^{-1}P(s)$. This term can be written as:

$$f(t) = \frac{\ln(2)}{t} \sum_{i=1}^{n} V_i P\left(\frac{\ln(2)}{t}i\right) \tag{A.1}$$

where $P\left(\frac{\ln(2)}{t}i\right)$ is yielded after substituting $\left(\frac{\ln(2)}{t}i\right)$ term in $P(s)$, and t is desired time at which Laplace inversion is needed. Also, V_i term is computed as shown in Eq. (A.2).

$$V_i = (-1)^{\frac{n}{2}+1} \sum_{k=\frac{i+1}{2}}^{\min\left(i,\frac{n}{2}\right)} \frac{k^{\frac{n}{2}}(2k)!}{\left(\frac{n}{2}-k\right)!k!(k-1)!(i-k)!(2k-i)!} \tag{A.2}$$

© The Author(s), under exclusive license to Springer Nature Switzerland AG 2022
A. Taghavinejad et al., *Unconventional Reservoirs: Rate and Pressure Transient Analysis Techniques*, SpringerBriefs in Petroleum Geoscience & Engineering,
https://doi.org/10.1007/978-3-030-82837-0

In this method, n is an arbitrary positive integer which should be determined by try and error. By increasing n value, until reaching an optimum number (the best selection for the computations), accuracy of Stehfest method increases as well. After this optimum value of n, by increasing this value accuracy will be reduced due to truncation error.

Appendix B

Based on the model developed by [3] for flow in shale gas reservoir producing with a multi-stage fractured horizontal well (MFHW)—which its primary modeling presented in Chap. 2, its semi-analytical solution for PTA and RTA applications presented in Chaps. 3 and 4, respectively—final Laplace space PTA solution for grids considered in each stage of hydraulic fractures (HFs) as depicted in Fig. 3.14, is as matrix equation written as Eq. 3.47. In this appendix of the book, MATLAB coding of plotting PTA type curves of the mentioned model is presented through Figs. B.1, B.2, B.3, B.4, and B.5.

As it can be seen in Figs. B.1 and B.2, rows 13–41 of this script define input parameters of the well, HFs, and reservoir. Also, other parameters such as inter-porosity flow parameters (inter-porosity flow coefficient, λ, fracture storativity, ω_f, and desorption storativity, ω_d) are calculated using input parameters through rows 43–61.

According to Fig. B.3, within rows 63–89, location of the grids on HFs are defined as longitudinal values in x and y directions.

As can be tracked in Fig. B.4, by defining time range with 60 points between $10^{-6.2}$ to $10^{3.5}$ with logarithmic spacing in row 91, Laplace inversion using Stehfest method is applied for solving matrix equation (Eq. 3.47) in each time-step (each of 60 points in the time range). Hence, by inserting a "for" loop in the code for dimensionless time values (row 92), a "for" loop for Stehfest method initialized as well (row 93). The "for" loop of Stehfest method is quantified within 1 to n (Stehfest number). So, by starting the "for" loop of Stehfest method, for each of 1 to n cases, matrix equation (Eq. 3.47) is solved, and by reaching the last case n (row 122), dimensionless pseudo-pressure (solution) is computed for the time value. Now, by iterating the loop for all dimensionless time values, dimensionless pseudo-pressure for the next time-steps will be computed.

Ultimately, taking advantage of codes which is written through rows 127–149 (Fig. B.5), PTA type curve (pseudo-pressure and pseudo-pressure derivative versus time in dimensionless form) for this well-reservoir model is plotted (Fig. B.6).

© The Author(s), under exclusive license to Springer Nature Switzerland AG 2022
A. Taghavinejad et al., *Unconventional Reservoirs: Rate and Pressure Transient Analysis Techniques*, SpringerBriefs in Petroleum Geoscience & Engineering,
https://doi.org/10.1007/978-3-030-82837-0

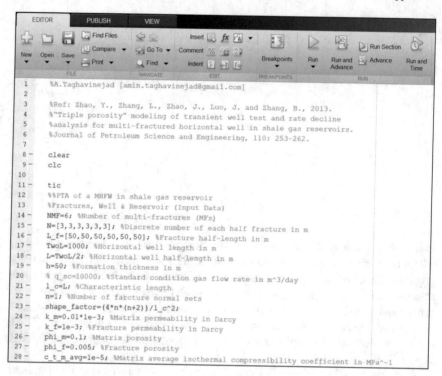

Fig. B.1 MATLAB script of PTA type curve plotting for an MFHW in shale gas reservoir (Part 1)

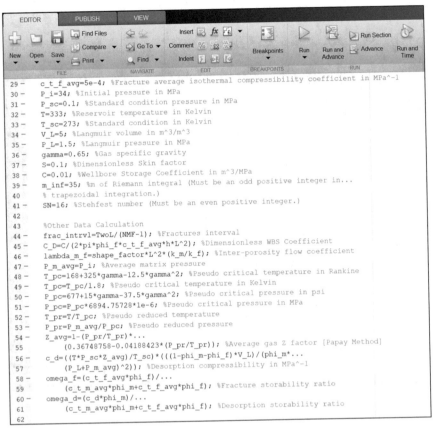

```
29 -  c_t_f_avg=5e-4; %Fracture average isothermal compressibility coefficient in MPa^-1
30 -  P_i=34; %Initial pressure in MPa
31 -  P_sc=0.1; %Standard condition pressure in MPa
32 -  T=333; %Reservoir temperature in Kelvin
33 -  T_sc=273; %Standard condition in Kelvin
34 -  V_L=5; %Langmuir volume in m^3/m^3
35 -  P_L=1.5; %Langmuir pressure in MPa
36 -  gamma=0.65; %Gas specific gravity
37 -  S=0.1; %Dimensionless Skin factor
38 -  C=0.01; %Wellbore Storage Coefficient in m^3/MPa
39 -  m_inf=35; %m of Riemann integral (Must be an odd positive integer in...
40    % trapezoidal integration.)
41 -  SN=16; %Stehfest number (Must be an even positive integer.)
42
43    %Other Data Calculation
44 -  frac_intrvl=TwoL/(NMF-1); %Fractures interval
45 -  C_D=C/(2*pi*phi_f*c_t_f_avg*h*L^2); %Dimensionless WBS Coefficient
46 -  lambda_m_f=shape_factor*L^2*(k_m/k_f); %Inter-porosity flow coefficient
47 -  P_m_avg=P_i; %Average matrix pressure
48 -  T_pc=168+325*gamma-12.5*gamma^2; %Pseudo critical temperature in Rankine
49 -  T_pc=T_pc/1.8; %Pseudo critical temperature in Kelvin
50 -  P_pc=677+15*gamma-37.5*gamma^2; %Pseudo critical pressure in psi
51 -  P_pc=P_pc*6894.75728*1e-6; %Pseudo critical pressure in MPa
52 -  T_pr=T/T_pc; %Pseudo reduced temperature
53 -  P_pr=P_m_avg/P_pc; %Pseudo reduced pressure
54 -  Z_avg=1-(P_pr/T_pr)*...
55       (0.36748758-0.04188423*(P_pr/T_pr)); %Average gas Z factor [Papay Method]
56 -  c_d=((T*P_sc*Z_avg)/T_sc)*(((1-phi_m-phi_f)*V_L)/(phi_m*...
57       (P_L+P_m_avg)^2)); %Desorption compressibility in MPa^-1
58 -  omega_f=(c_t_f_avg*phi_f)/...
59       (c_t_m_avg*phi_m+c_t_f_avg*phi_f); %Fracture storability ratio
60 -  omega_d=(c_d*phi_m)/...
61       (c_t_m_avg*phi_m+c_t_f_avg*phi_f); %Desorption storability ratio
62
```

Fig. B.2 MATLAB script of PTA type curve plotting for an MFHW in shale gas reservoir (Part 2)

```
63        %Coordination definition
64 -   for m=1:NMF
65 -       for k=1:NMF
66 -           for g=N(k)|
67 -               if k==m
68 -                   Delta_L_f_i=L_f(m)/N(m);
69 -                   Delta_L_f_D_i=Delta_L_f_i/L;
70 -                   for i_num=1:2*N(m)
71 -                       x_i=-(2*N(m)-(2*i_num-1))/(2*N(m))*L_f(m);
72 -                       x_D_i=x_i/L;
73 -                       y_D_i=((m-1)*frac_intrvl)/L;
74 -                       if m==1
75 -                           Delta_L_f(i_num)=Delta_L_f_i;
76 -                           Delta_L_f_D(i_num)=Delta_L_f_D_i;
77 -                           x_D(i_num)=x_D_i;
78 -                           y_D(i_num)=y_D_i;
79 -                       else
80 -                           Delta_L_f(i_num+2*sum(N(1:m-1)))=Delta_L_f_i;
81 -                           Delta_L_f_D(i_num+2*sum(N(1:m-1)))=Delta_L_f_D_i;
82 -                           x_D(i_num+2*sum(N(1:m-1)))=x_D_i;
83 -                           y_D(i_num+2*sum(N(1:m-1)))=y_D_i;
84 -                       end
85 -                   end
86 -               end
87 -           end
88 -       end
89 -   end
90
```

Fig. B.3 MATLAB script of PTA type curve plotting for an MFHW in shale gas reservoir (Part 3)

```
63        %Coordination definition
64 -   ☐for m=1:NMF
65 -        for k=1:NMF
66 -            for g=N(k)|
67 -                if k==m
68 -                    Delta_L_f_i=L_f(m)/N(m);
69 -                    Delta_L_f_D_i=Delta_L_f_i/L;
70 -                    for i_num=1:2*N(m)
71 -                        x_i=-(2*N(m)-(2*i_num-1))/(2*N(m))*L_f(m);
72 -                        x_D_i=x_i/L;
73 -                        y_D_i=((m-1)*frac_intrvl)/L;
74 -                        if m==1
75 -                        Delta_L_f(i_num)=Delta_L_f_i;
76 -                        Delta_L_f_D(i_num)=Delta_L_f_D_i;
77 -                        x_D(i_num)=x_D_i;
78 -                        y_D(i_num)=y_D_i;
79 -                        else
80 -                        Delta_L_f(i_num+2*sum(N(1:m-1)))=Delta_L_f_i;
81 -                        Delta_L_f_D(i_num+2*sum(N(1:m-1)))=Delta_L_f_D_i;
82 -                        x_D(i_num+2*sum(N(1:m-1)))=x_D_i;
83 -                        y_D(i_num+2*sum(N(1:m-1)))=y_D_i;
84 -                        end
85 -                    end
86 -                end
87 -            end
88 -        end
89 -   end
90
```

Fig. B.4 MATLAB script of PTA type curve plotting for an MFHW in shale gas reservoir (Part 4)

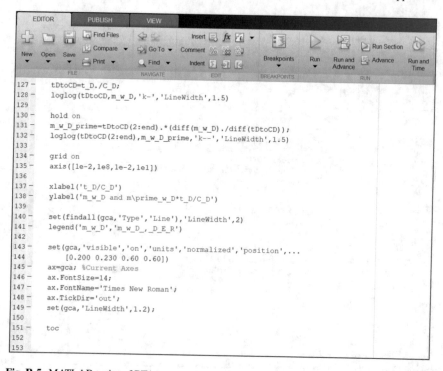

Fig. B.5 MATLAB script of PTA type curve plotting for an MFHW in shale gas reservoir (Part 5)

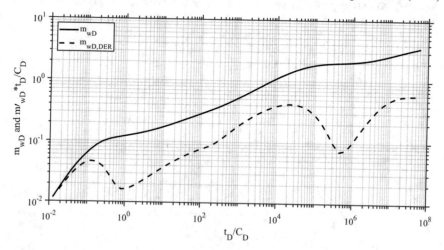

Fig. B.6 PTA type curve for an MFHW in shale gas reservoir

Appendix C

The MATLAB script for plotting RTA type curve for an MFHW in a shale gas reservoir is generally similar to the PTA type curve plotting code which was explained in Appendix B. The type curve which is explained in this appendix of the book is also based on the model presented by [3].

Regarding what said in Chap. 5, by taking advantage of the principle proposed by [2], to reach the rate solution under production conditions of constant bottom-hole flowing pressure, it is enough to add Laplace space dimensionless rate expression (Eq. 4.37) in the MATLAB script reviewed in Appendix B. Figure C.1 shows the part of MATLAB script of this case in which mentioned command is added in row 112. Then, in the next rows, Stehfest method should be applied on dimensionless rate instead of dimensionless pseudo-pressure (rows 113–125). Also, dimensionless rate derivative has to be used instead of dimensionless pseudo-pressure derivative as it is written in row 132. For rest of the code, the titles and legend of the plot is edited for RTA type curve case.

© The Author(s), under exclusive license to Springer Nature Switzerland AG 2022
A. Taghavinejad et al., *Unconventional Reservoirs: Rate and Pressure Transient Analysis Techniques*, SpringerBriefs in Petroleum Geoscience & Engineering,
https://doi.org/10.1007/978-3-030-82837-0

```
111 -    m_w_D_bar(sn)=(S+X(end)*s)/(s+C_D*s^2*(S+X(end)*s));
112 -    q_D_bar(sn)=1/(s^2*m_w_D_bar(sn));
113 -    S_V=0;
114 -        for k_S=floor((sn+1)/2):min(sn,SN/2)
115 -            S_V=S_V+(((k_S^(SN/2))*factorial(2*k_S))/(factorial(SN/2-k_S)...
116               *factorial(k_S)*factorial(k_S-1)*factorial(sn-k_S)*...
117               factorial(2*k_S-sn)));
118 -        end
119 -        V(sn)=((-1)^(SN/2+sn))*S_V;
120 -    end
121 -    S_q_D=sum(V.*q_D_bar);
122 -    t_D_numb=find(t_D_now==t_D);
123 -    q_D(t_D_numb)=(log(2)./t_D_now)*S_q_D;
124 -    fprintf('- q_D for the %g/%g of t_D values is calculated.\n',t_D_numb,...
125         length(t_D))
126 -  end
127
128 -    tDtoCD=t_D./C_D;
129 -    loglog(tDtoCD,q_D,'k-','LineWidth',1.5)
130
131 -    hold on
132 -    q_D_prime=tDtoCD(2:end).*(-diff(q_D)./diff(tDtoCD));
133 -    loglog(tDtoCD(2:end),q_D_prime,'k--','LineWidth',1.5)
134
135 -    grid on
136 -    axis([1e-2,1e8,1e-2,1e1])
137
138 -    xlabel('t_D/C_D')
139 -    ylabel('q_D and q\prime_D*t_D/C_D')
140
141 -    set(findall(gca,'Type','Line'),'LineWidth',2)
142 -    legend('q_D','q_D_,_D_E_R')
143
144 -    set(gca,'visible','on','units','normalized','position',...
145         [0.200 0.230 0.60 0.60])
146 -    ax=gca; %Current Axes
147 -    ax.FontSize=14;
148 -    ax.FontName='Times New Roman';
149 -    ax.TickDir='out';
150 -    set(gca,'LineWidth',1.2);
151
152 -    toc
153
```

Fig. C.1 Last part of the MATLAB script of RTA type curve plotting for an MFHW in shale gas reservoir

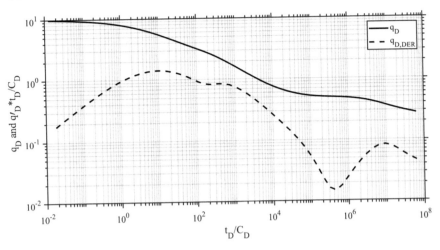

Fig. C.2 RTA type curve for an MFHW in shale gas reservoir

Figure C.2 demonstrates the plotted RTA type curve (dimensionless rate and dimensionless rate derivative versus dimensionless time) of this model

References

1. Stehfest H (1970) Numerical inversion of Laplace transform. Commun ACM 13:47–49. https://doi.org/10.1145/361953.361969
2. van Everdingen AF, Hurst W (1949) The application of the Laplace transformation to flow problems in reservoirs. J Pet Technol 1:305–324. https://doi.org/10.2118/949305-G
3. Zhao Y, Zhang L, Zhao J et al (2013) "Triple porosity" modeling of transient well test and rate decline analysis for multi-fractured horizontal well in shale gas reservoirs. J Pet Sci Eng 110:253–262

Printed in the United States
by Baker & Taylor Publisher Services